FOUNDATIONS 1999

Ecological Studies

Analysis and Synthesis

Edited by

W. D. Billings, Durham (USA) F. Golley, Athens (USA)
O. L. Lange, Würzburg (FRG) J. S. Olson, Oak Ridge (USA)

Volume 22

Robert Guderian

Air Pollution

Phytotoxicity of Acidic Gases and Its Significance in Air Pollution Control

Translated from the German by C. Jeffrey Brandt, M. Sc.

With 40 Figures, 4 in Color

Springer Verlag Berlin Heidelberg New York 1977

Dr. ROBERT GUDERIAN
Landesanstalt für Immissions- und Bodennutzungsschutz
des Landes Nordrhein-Westfalen
Wallneyer Str. 6, 4300 Essen/FRG

ISBN 3-540-08030-9 Springer-Verlag Berlin · Heidelberg · New York
ISBN 0-387-08030-9 Springer-Verlag New York · Heidelberg · Berlin

Library of Congress Cataloging in Publication Data. Guderian, Robert. Air pollution. (Ecological studies; v. 22) Bibliography: p. 1. Plants, Effect of air pollution on. 2. Plant indicators. 3. Air —Pollution — Measurement. I. Title. II. Series. QK751.G77. 581.5'222. 76-50626.

Typesetting, printing, and binding: Brühlsche Universitätsdruckerei, Gießen

2131/3130-543210

Introduction

Emissions of gaseous air pollutants have increased in the last years in spite of increased controls and concern for air quality. Predictions of future development also indicate that a further increase in emissions must be expected. From an extensive analysis of fuel use in conventional power plants in industry and for domestic heating, Brocke and Schade (1971) and Schade (1975) predict that sulfur dioxide (SO_2) emissions in the Federal Republic of Germany will increase from 3.5 million t in 1969, over 4.2 million t in 1973, to 4.6 million t in 1980. Rasch (1971) predicts that emission of hydrogen chloride (HCl) from burning of wastes will increase from a present 8000 t/year to about 100000 t in 1980. Emission of gaseous fluoride compounds, in North Rhine Westphalia alone, are expected to increase from 7500 t in 1969 to 8800 t in 1985 (MAGS, 1972). Similar predictions have also been made in the USA (Heggestadt and Heck, 1971). A doubling of SO_2 emissions from oil and particularly coal-fired power plants is expected between 1960 and 1980 (Wood, 1968; Lewis et al., 1974).

When it is considered that total control of air pollutants is technically and especially economically impossible, it is important that, in the future, emissions are controlled within a technical and economic framework to such an extent that ambient pollutant concentrations near the ground present no hazard to man or his environment (BImSchG, 1974).

In order to evaluate and set such limits for allowable pollutant concentrations, knowledge of the quantitative relationships between ambient pollutant concentrations and the object affected is necessary. Since several effective components occur at the same time, studies on the effects of pollutants in combination are essential.

Compared with man, animals, or materials, plants respond very sensitively to widely distributed pollutants, such as sulfur dioxide, hydrogen fluoride (HF) and hydrogen chloride. Extensive loss to agriculture and lasting changes in natural ecosystems are the result. Studies on the effects of air pollution on vegetation, therefore, provide an important basis, particularly for preventive measures in air pollution control.

Such studies require a characterization of the pollutant situation that is based on effects of pollutants. Also necessary is an evaluation, based on economical and ecological criteria, of these effects on the intended use of particular plant species. Responses of plants to pollutants are not only primarily dependent on pollutant concentration and exposure time, but also on the amount of pollutant absorbed by the plant per unit of time (Guderian, 1970). The rate of uptake varies just as much as the species and variety-specific resistance during plant development and as external growth factors. The determination of representative dose-response relationships, therefore, is quite complex and can only be accomplished through consideration of an extensive set of conditions.

In spite of protective measures through environmental and industrial planning, as well as the setting of standards for allowable pollutant concentrations, plant injury from atmospheric pollutants still occurs. Ascertaining the cause of injury is decisive for the selection of controls for the pollutant source and protective measures at the growing site, and for solving such cases where injury is present. Symptoms of air pollutant injury, such as necrosis, chlorosis, or changes in growth habit are not specific enough to serve alone as positive identification of pollutant effects (Dörries, 1932; Massay, 1952; Guderian and van Haut, 1970).

In such cases, which are based mainly on inductive reasoning, other criteria, such as chemical analysis of plant material for accumulated pollutants, must also be used. The diagnostic value of this method rests primarily on the amount of a particular component absorbed from the air compared with the variability of the natural content of this element in the plant. Studies on the dependence between pollutant accumulation in plant organs and internal and external growth factors are necessary for determining the limits and possibilities for use of this diagnostic method, as well as for the use of plants as biologic indicators in polluted areas.

Because of their complexity, relationships between pollutant concentrations and their effects cannot be studied in a single series of experiments. Studies near pollutant sources, as well as in fumigation chambers in the field and laboratory are required.

For the detection and interpretation of plant responses, chemical analysis and determination of quality and yield, as well as gas-exchange measurements and electronmicroscopic studies are used. Through a presentation of the author's personal studies and an evaluation of the available literature, dose-response relationships and their connection with internal and external growth factors are described. The requirements and problems of such studies are also presented.

Using sulfur dioxide, hydrogen fluoride, and hydrogen chloride as examples of important pollutants in Europe, a discussion of the characteristic effects of these pollutants and their influence on the quantitative relationships mentioned is presented. The last section deals with the importance of these experimental results for the development of preventive and augmentative measures in pollution control.

Essen, January 1977 ROBERT GUDERIAN

Contents

1. Materials and Methods

Considerable experimental work is required to ascertain representative quantitative relationships between ambient air pollutants and their effects on plants. This is a result not only of the necessity for monitoring the pollutant concentrations and the determination of various plant responses that relate to the useful value of the plant, but also of specific experimental requirements that take into account the susceptibility of the plant material.

1.1 Apparatus for the Experimental Determination of Air Pollution Effects

Three different types of facility should be considered for studies on the effects of air pollutants on vegetation. These include investigations in the laboratory and in controlled-environment chambers, under glass or in plastic greenhouses in the field, and investigation of test plants near sources of pollution. Each of these techniques is useful for specific experimental problems (Guderian, 1966a). For the investigation of complex problems, such as the dose–response relationships studied here, experiments should be carried out with each of these methods.

The description of the procedures used in the author's personal experiments includes a comparative discussion of the technical qualities of the three methods. This is intended to show which methods are optimally suited for special studies and which criteria should be chosen for evaluation of experimental data in regard to their use in air pollution control.

1.1.1 Field Study Near a Sulfur Dioxide Source

Various plants were set out in pots in the area near an iron ore smelter to determine phytotoxic sulfur dioxide concentrations.

Six stations were set up at different distances from the smelter on sites having different SO_2 concentrations but comparable climatic conditions (Guderian, 1960; Guderian and Stratmann, 1962). Five of the six stations were within effective range of the pollutant source and the sixth served as the control. The SO_2 concentration was monitored continually with Ultragas III instruments, which use the conductivity method. Table 1 shows the characteristic pollutant conditions at the single stations.

Fig. 1. Experiment station near an SO_2 source

Table 1. Technical characteristics of the experiment stations in the field experiment near an SO_2 source

Station	Distance from the source (in m)	Degree of injury of plants	Exposure time t_i in % of the monitoring time t_m	Average concentrations in ppm[a]	
				c_i during exposure time (t_i)	c_m during monitoring time (t_m)
I	325	Very severe to total injury	24.1	0.59	0.141
II	600	Very severe injury	18.9	0.44	0.083
III	725	Severe injury	14.7	0.34	0.051
IV	1350	Slight injury	7.6	0.26	0.020
V	1900	Slight injury to very sensitive plants	4.3	0.22	0.010
VI	6000 Control	No injury	No measurable SO_2		

[a] Average of two monitoring times, each from 1 April to 31 October.

Fig. 2. Fumigation chambers in the field for the study of air pollution effects on vegetation

The mean SO_2 concentrations, $\bar{c}_m$ and $\bar{c}_i$, during monitoring and exposure times t_m and t_i, respectively, were computed with the following formulas:

$$\bar{c}_m = \frac{[\Sigma(c \cdot \Delta t)]\, c \geqq 0.10\,\text{ppm}}{t_m}$$

$$\bar{c}_i = \frac{[\Sigma(c \cdot \Delta t)\, c \geqq 0.10\ \text{ppm}}{t_i}\,.$$

The monitoring time, t_m, is essentially equal to the exposure time of the test plants. The exposure time, t_i, was calculated by summing all time intervals, $\Delta t = 10$ min, with a mean SO_2 concentration greater than or equal to 0.10 ppm. With increasing distance from the source, the exposure time decreased, so that, based on the defining formula, the arithmetic mean, $\bar{c}_m$, over monitoring time t_m is very small.

1.1.2 Fumigation Experiments in the Field

In field experiments near pollutant sources, the varying pollutant conditions must be accepted. On the other hand, use of small glass or plastic greenhouses at field stations allows fumigation with constant pollutant concentrations—within a specific deviation range (Fig. 2).

Fumigation chambers, consisting of a frame covered with sheets of Mylar polyester, are supplied with air/pollutant mixtures through underground plastic ducts (Fig. 3.15) from a climatically controlled machine room. To avoid condensation and absorption of the pollutant in the ducts, moisture is removed from the carrier air by freeze drying (Fig. 3.1). The air/pollutant mixture is diluted to fumi-

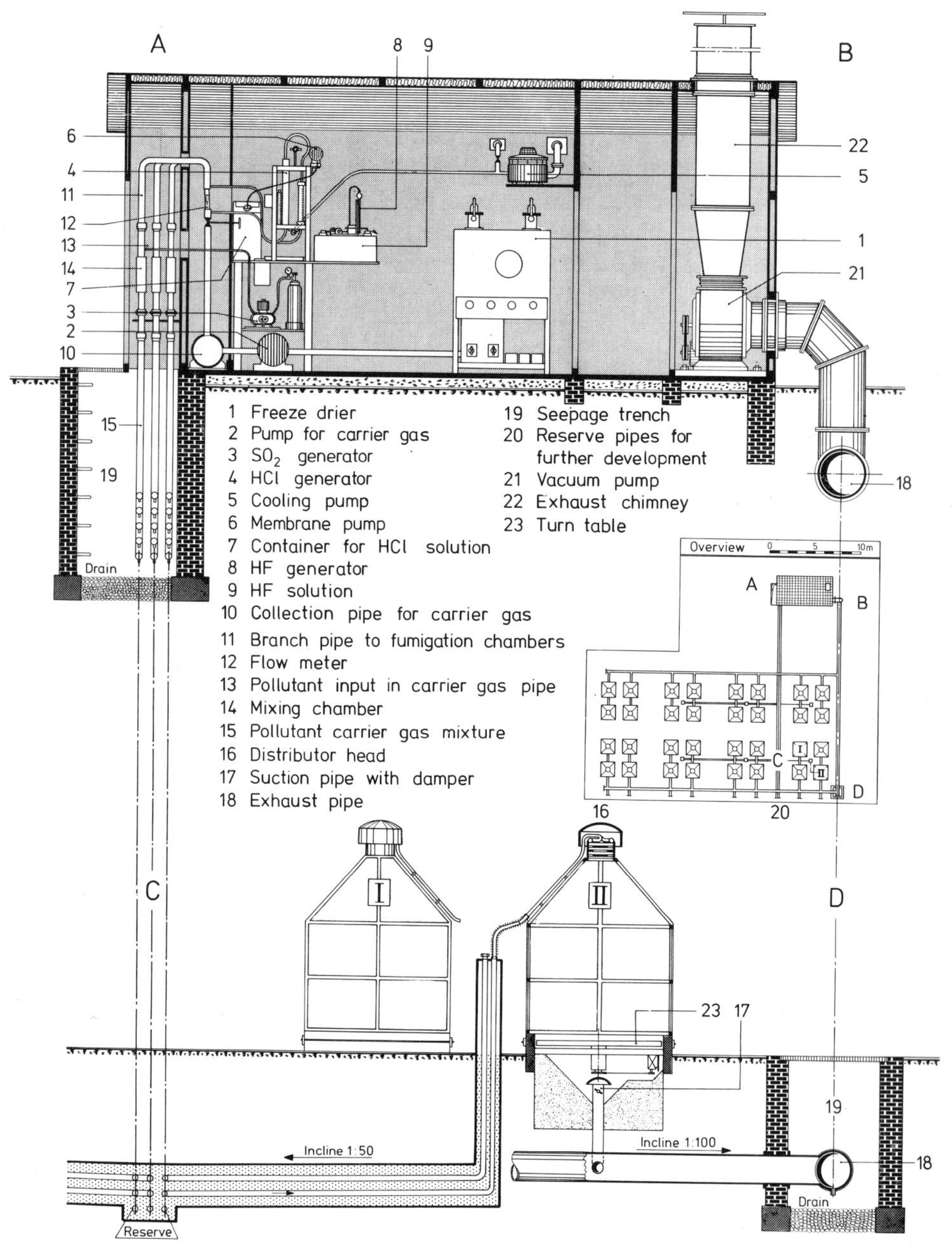

Fig. 3. Schematic drawing of the field station

gation concentrations in the head manifold (Fig. 3.16) on the chamber roof with outside air sucked into the chamber through activated charcoal filters (Fig. 3.17). The air/pollutant mixture is then pumped underground (Fig. 3.18) and, after washing, through an exhaust duct (Fig. 3.22).

The pollutant concentration in the chamber is regulated by the amount of pollutant generated and the amount of make-up air used for dilution. Make-up air can be regulated by a continuously variable damper in the incoming air stream. With a high rate of air exchange—usually 80–100 times/h—variations in temperature and humidity in the chambers follow those of the outside air. Turntables (Fig. 3.23) ensure that all test plants are exposed to the same pollutant and climatic conditions.

Generation of sulfur dioxide is carried out with mixing pumps (Fig. 3.3). For the generation of hydrogen fluoride, dried air is passed over dilute hydrofluoric acid in a temperature-controlled plastic container. The carrier air is loaded with hydrogen fluoride according to the partial pressure of HF over the hydrofluoric acid solution. Specific HF concentrations can be produced through variation of temperature, hydrofluoric acid concentration, and/or air stream. Hydrogen chloride concentrations are generated by the votalilization of specific amounts of aqueous HCl in heated tubing. The HCl solution is pumped by peristaltic pumps (Fig. 3.6).

Monitoring of the actual concentration in the chamber is carried out with physical–chemical methods described in Section 1.2.

1.1.3 Fumigation Experiments in Controlled-environment Chambers

As a result of the extremely good control over all conditions, fumigation experiments in controlled-environment chambers allow the investigation of the effects of single factors or a complex of factors. Test plants are placed in a 3 m^3 test chamber in which temperature and humidity are regulated through wall temperature of the chamber and exchange—up to 200 m^3/h—of climatized air (Arndt et al., 1973). Two xenon lamps, each with an output of 6 KW, provide 20000 lux illumination at the chamber floor.

An apparatus which can be rolled into the controlled-environment chambers was developed (Guderian and Thiel, 1973) for the study of combined effects of several simultaneously or alternately occurring pollutants. The test plants are fumigated in four 30 l cylindrical plexiglas chambers (Fig. 4).

Air is drawn through the chambers by vacuum pumps and is regulated by valves and flow meters (Fig. 5). Pollutants are introduced at the inlets of the chambers; SO_2 by means of several Wösthoff pumps connected in series and HCl by means of the capillary system developed by Hartkamp (1973).

In experiments on the combined effects of SO_2 and HCl, one chamber contained only SO_2, a second only HCl, and a third a combination of the two pollutants. The nonfumigated test plants in the fourth chamber served as the control. The construction of the apparatus, together with the pollutant production systems, ensured constant pollutant concentrations over any time period.

The CO_2 and O_2 gas exchange of the test plants during fumigation can be monitored respectively with Unor or Oxygor instruments and transpiration can

Fig. 4. Experimental apparatus for determining effects of pollutants in combination. *Left front:* Wösthoff instrument for monitoring SO_2 and HCl concentrations; *left rear:* UNOR infrared gas analyzer for monitoring CO_2 content of the air

be measured with a specially modified Uras instrument. Solenoid valves at the inlet and outlet of each chamber can be activated in any desired order. An air stream, regulated with a flow meter, flows from the monitor point to the solenoid valve. Thus, for example with the Unor, flushing of the tubing can be accomplished in about 10 sec and the actual instantaneous CO_2 concentration in the chamber is recorded. The switching time of the solenoid valves can be set for 1 min, so that, within 6 min, all six monitor points in the system can be recorded.

1.1.4 Comparison of Experimental Methods

The experimental systems described in the preceding sections serve as representative examples of the three experimental methods mentioned. Their suitability for the investigation of particular problems depends largely on the type and amount of pollutant, as well as on climatic factors, as shown in Table 2.

Studies on the effects on normally predisposed test plants under varying pollutant and climatic conditions are possible in field experiments near single sources or in areas under pollutant stress from several sources. Dose–response relationships, determined in such experiments, are the basis for defining potential areas of hazard to vegetation, especially because long-term effects on perennial cultures and other ecosystems can be determined in this way. The experimental investment

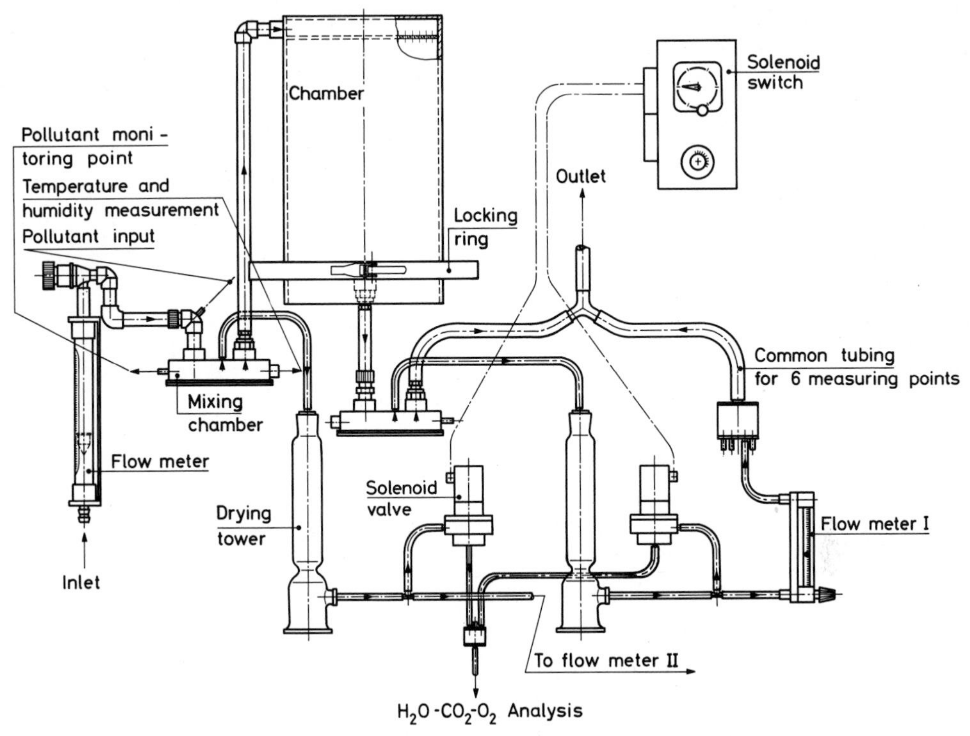

Fig. 5. Single chamber with regulation and supply mechanisms

is, of course, quite large, not just because of difficulties in determining the prevailing pollutant stress. Such experiments, as a result of the many variables, are always multifactorial and determination of the effect of single factors on the total result is seldom, if ever, possible.

In growth chamber experiments, as opposed to open-field studies, pollutant concentrations, as well as climatic factors, can be regulated so that the importance of single factors, or a complex of factors can be reproducibly determined. Depending on the type of test chamber, factors that determine growth deviate more or less from those in the field. Long-term fumigations to determine quantitative relationships are, therefore, problematic. Long-term experiments on trees and shrubs are also impossible due to spatial requirements.

In their technical requirements and in regard to their usefulness, fumigation experiments in greenhouses, under conditions similar to those in the field, lay between the two methods described above.

1.2 Air Analyses

Sulfur dioxide and hydrogen chloride concentrations in field studies and in fumigation experiments in greenhouses and growth chambers were continuously monitored with Ultragas-III instruments manufactured by Wösthoff. These in-

Table 2. Comparison of characteristics of each experimental design

Characteristic factors	Field experiments in polluted areas	Fumigation experiments under glass and plastic chambers in the field	Fumigation experiments in growth chambers
Pollutant supply	Constant variation of pollutant concentration and exposure time	Limited regulation of pollutant concentration and exposure time	Complete regulation of pollutant concentration and exposure time
Climatic conditions	Natural variations	Slightly altered natural conditions	Controlled conditions
Plant sensitivity	Normal predisposition	Normal predisposition under certain conditions	Limited normal to abnormal predisposition, depending on climatic conditions
Evaluation	Always multifactorial, experiments not reproducible because of generally large time investment, study of complex effects of continuously changing pollutant and climatic conditions, long-term studies of natural and altered ecosystems for determining dose – response relationships	Limited multifactorial, certain experiments reproducible, limited suitability for study of single factors, such as concentration and exposure time, as well as stage of development and leaf age, limited suitability for study of dose-response relationships, experiments possible over a period of several years	Control of pollutant and climatic conditions, experiments are reproducible, studies on single factors for plant reactions, determination of quantitative relations between pollutant and effects as long as it is not dependent on normal plant predisposition, short-term experiments up to several months

struments measure the change in conductivity of a specific reaction solution caused by absorption of the pollutant. Hydrogen fluoride concentrations were monitored with the discontinuous method developed by Buck and Stratmann (1965). This method, which employs bicarbonate-coated silver beads in a glass tube, measures essentially only gaseous fluoride, the major causative agent in fluoride injury to vegetation. The analysis of F^- ions in the eluate was carried out with an ion-specific electrode (Reusmann and Westphalen, 1969).

1.3 Test Plants and Soils

Experiments were carried out with numerous plant species important in agriculture, reflecting the goal of defining hazards to vegetation. A detailed description is included in the discussion of experimental results. Various kinds of container (plastic pots, containers developed by Mitscherlich, or those from Kick/Grosse-Bauckmann, wooden tubs), with a capacity up to 60 l, were used, dependent upon plant type and experimental requirements. Plants were also set out in beds under fumigation chambers and samples were taken from natural vegetation within and outside polluted areas.

In order to guarantee normal growth conditions, special soil mixtures were developed for the various cultures. Mixing of soil components and inorganic fertilizers was done in a mixing machine. Samples were taken to determine nutrient levels and to check uniformity of the mixture (Guderian, 1960). Nutrient levels were such that average to good growth was assured.

1.4 Determination of Plant Responses

1.4.1 Growth, Yield, and Quality

Changes in growth, yield, and quality serve as a basis for ascertaining injury limits, as these criteria give clear information on the useful value of the particular plant species. Fresh weight and dry matter production were determined. On trees and shrubs, effects on growth were determined through branch and lateral growth measurements, as well as through yield determinations.

Evaluation of external appearance and analysis for particular key substances gave an indication of the reduction in quality. In other experiments, information on effects on seed quality and reproduction was obtained (Guderian and Stratmann, 1962, 1968). Degree of foliar injury is given either in percent of necrotic area or based on the following table:

1. Very slight necrosis or very slight chlorosis
2. Slight necrosis or slight chlorosis
3. Medium necrosis or medium chlorosis
4. Severe necrosis or severe chlorosis
5. Very severe necrosis or very severe chlorosis.

1.4.2 Gas-exchange Measurements

Effects on gas-exchange were determined through measurements of apparent photosynthesis of either entire plants in plexiglas chambers (see Sect. 1.1.3) or of single leaves in special leaf chambers (Guderian, 1970). Both absolute and difference (Egle, 1960) measurements were carried out with infrared gas analyzers (UNOR and 15A manufactured by Maihak, and Beckmann, Inc. respectively). The absolute CO_2 exchange/unit of time is calculated by the product of the difference in CO_2 concentration in incoming and outgoing air and the air exchange rate, and can be given in mg CO_2/dm^2 leaf area · unit of time.

Photosynthesis measurements were carried out in growth chambers and in the field. The air exchange rate was chosen so as to avoid a CO_2 deficit and to obtain significant measurements. These requirements were met with exchange rates between 40 and 200 times/h, depending on the relation of leaf area to chamber volume, as well as consideration of metabolic activity and climatic conditions.

Pots with decapitated plants or pots in which the soil was isolated with plastic bags were also placed in the chambers during the experiment in order to measure CO_2 production of the soil for the determination of net photosynthesis.

1.4.3 Chemical Analysis of Plant Material

To measure the accumulation of pollutants in plants, chemical analyses for S, F, and Cl in the plant material were carried out. Chloride and sulfur were analyzed with the method developed by Reusmann and Westphalen (1976). About 500 mg dried plant material is combusted with sodium peroxide in a Wurzschmitt bomb. After dissolving the cooled product in water and acidification with formic acid, analysis is carried out with an Autoanalyzer. For the analysis of sulfur, the solution is mixed with a reducing solution consisting of formic acid, hypophosphoric acid, and hydriodic acid. This solution, together with nitrogen, is passed through a coil, which is heated to 110° C, and reduction to hydrogen sulfide takes place. An acidic molybdenum solution is added to the nitrogen/hydrogen sulfide mixture and the amount of molybdenum blue produced is measured photometrically at 580 nm. Transmission is recorded.

For the analysis of chloride, the prepared sample is mixed with a nitric acid solution containing iron III nitrate, mercuric thiocyanate, and methanol. The red color complex, formed in the presence of chloride ions, is measured photometrically.

In the determination of fluoride, 1 g dried plant material is ashed, dissolved in NaOH, and, after addition of a citrate buffer, the fluoride ion content is measured with an ion specific electrode (Reusmann and Westphalen, 1969) in an automatic system (Buck and Reusmann, 1971).

1.4.4 Electron-microscopic Analysis of Ultrastructure

Objects to be studied with the electron microscope were fixed in a $KMnO_4$ solution, buffered with 0.6% veronal acetate (Luft, 1956), a 1% OsO_4 solution (Palade, 1952) or in 6.25% glutaraldehyde (Sabatini et al., 1963). After fixation in the aldehyde, which took 1.5 to 2 h, contrast was established by placing the object in 1% OsO_4 for 2 h. Fixation times with the other methods lay between 3 and 10 h. After dehydration in an alcohol or acetone dilution series, the major portion of lipid materials remained (Masuch et al., 1973). Imbedding was carried out in four steps in EPON epoxy (Spurr, 1961). Polymerization took place in drying ovens; 24 h at 35° C, 24 h at 45° C, and 48 h at 70° C. Ultra-thin sections were prepared with a Porter-Blum microtome. Contrast was established in 0.5–3 min with lead citrate (Reynolds, 1963). Analysis of the sections was carried out on an Elmiskop I electron microscope manufactured by Siemens.

For quantitative analysis, plastids were measured on the photographs with a planometer and observed components were related to 100 μm^2 plastid-area—that is, the plasma area surrounded by the double membrane. No series sections were used to ensure representative results. The results, as Schnepf (1963) also stresses, are not absolute values. Comparisons with nonfumigated objects, however, allow the recognition of structural changes due to the influence of air pollutants.

2. Experimental Analysis of the Effects of Gaseous Air Pollutants

The type and extent of effects of an air pollutant on vegetation depend on the amount of pollutant present and on the genotypically and environmentally determined plant resistance. Therefore, the following requirements are necessary for establishing quantitative relationships between pollutant and effects, relationships that serve as a basis for risk-prognoses and pollutant control measures. The extent of atmospheric pollutant load, in terms of concentration and exposure time, must first be determined by chemical–physical means. It is particularly important that the most effective component is monitored. For example, with fluoride the most important components are the water-soluble gaseous fluoride compounds, such as hydrofluoric acid (HF), silicon tetrafluoride (SiF_4), and fluorosilicic acid (H_2SiF_6) (Buck and Stratmann, 1965; Guderian et al., 1969; Weinstein and Mandl, 1971).

The pollutant load, determined in a relatively short sampling time (Stratmann, 1963a; Zahn, 1963a), is defined by various effects criteria and compared with plant responses described in Section 2.1. These reactions range from disruption of biochemical reactions and microscopic and submicroscopic changes in cell organelles, through changes in morphology and habit, to effects on the entire organism, which are expressed as reduced growth, reduced quality, or even death of the plant (cf. discussion in Sect. 3.3)

The effect-response is composed of two processes; uptake of the pollutant and response of the plant to the incorporated pollutant. A discussion of the changes in disposition of the plant (Gäumann, 1951) under a given pollutant load and under influence of interacting internal and external growth factors is presented in the following sections.

2.1 Criteria for the Evaluation of Air Pollution Effects

As mentioned in the Introduction, the determination of limits of injury serves as a basis for ascertaining areas of hazard to vegetation and for setting allowable air pollutant limits based on vegetation effects. If vegetation, in its function as an economic object or as a part of healthy ecosystems, is to be protected, criteria for dose–response relationships must be chosen that yield information regarding the usefulness of the various plant species. In other words, the intended use of the

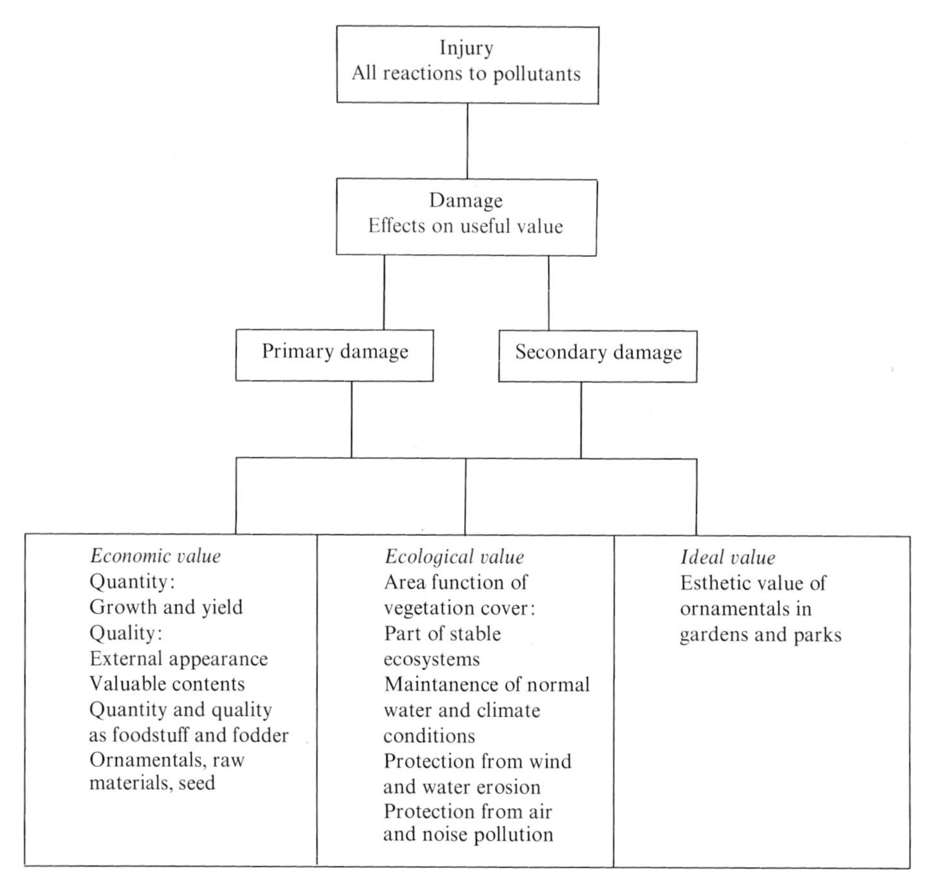

Fig. 6. Effect of air pollutants on the useful value of plants

plant determines the methods of analysis for pollutant effects, a condition which has been discussed earlier (Guderian et al., 1960). This division of air pollutant effects on vegetation into "injury" and "damage" has now become generally accepted (Brandt, 1962; Kisser et al., 1962; Adams, 1963; Manschinger, 1966; Wentzel, 1967; Heggestad and Heck, 1971). This definition also allows the comparison of effects on different crops, which was recognized by Zieger as early as 1953 as a special problem.

The term "injury" includes all plant responses that occur due to atmospheric pollution; reversible alterations in metabolism, as well as reductions in photosynthesis, foliar necrosis, leaf-fall, or growth reduction. The term "damage" includes all effects which reduce the intended value or use of the plant (Fig. 6). The useful value—determined by economic, ecologic, and esthetic values—can be reduced through effects on growth, yield or quality. Indirect damage can occur after accumulation of pollutants in the soil (MacIntire et al., 1958, Meran, 1960; Kozel and Maly, 1968).

Although much experimental information is available concerning this so-called primary injury, little is known about "secondary injury," which occurs from the effects of biotic and abiotic factors as a result of reduced resistance of the plant (Donaubauer, 1966). Examples of secondary injury include reduced frost resistance of winter grain crops (Guderian and Stratmann, 1962, 1968), of forest trees (Wentzel, 1956, 1965), and of fruit trees (Auersch, 1967), as well as the increased incidence of some parasites (Templin, 1962; Sierpinski, 1967; Bösener, 1969).

Effects on growth, yield and quality—a sometimes subjective factor—are the criteria most often used for determining economic damage to vegetation used in agriculture, forestry, and gardening. Together with effects on products used for food or fodder, or as raw materials, effects on seed quality, for example in potato (Guderian and Stratmann, 1962) or gladiolus (Spierings, 1963), as well as disruption of pollination in forest cultures with negative effects on natural reproduction (Pelz, 1963) are also important. Reduction in the quality-complex occurs not only through changes in external appearance, but also through alterations in the composition of metabolic products (Reckendorfer and Beran, 1931), through accumulation of pollutant compounds in plant material (Rosenberger, 1963; Rippel, 1972), as well as through changes in the composition of plant communities (Guderian, 1966b). Additional losses occur in agriculture due to the fact that certain sensitive species cannot be grown in polluted areas (Wentzel, 1963).

The ecologic value of vegetation, which is a function of area (Niesslein, 1966), is based on specific beneficial functions (Speidel, 1966) of the vegetation which include vegetation as a part of healthy agricultural and recreation areas, as necessary for maintenance of water and climatic conditions, for protection against wind and water erosion, and for reduction of stresses from atmospheric pollution (Hennebo, 1955; Bernatzky, 1968; Steubing and Klee, 1970) and noise pollution (Dreyhaupt, 1971; Meurers, 1972). These ecologic functions are disrupted over large areas (Materna, 1965b; Wentzel, 1967; Knabe, 1970a), as a reduction in the ability for regeneration leads to a reduction in strength of plant communities and, therefore, to alterations in biocoenoses, as well as to unstable ecosystems (Borgsdorf, 1960; Nikfeld, 1967; van Haut and Stratmann, 1970).

The esthetic value, for example of ornamentals in gardens and parks, is lessened primarily through foliar injury rather than through reduced growth. This type of damage has increased considerably as a result of extensive construction of industrial and residential areas, and leads more and more frequently to complaints (Hölte, 1972).

2.2 Influence of Concentration and Exposure Time

Experiments to determine the quantitative relationships between pollutant concentration in the air and effects on plants were already carried out in the last century. After Stöckhardt (1850, 1871), in a systematic series of experiments which were decisive in establishing pollution studies as a scientific discipline, was able to show that SO_2 causes injury to vegetation, details of the extent of injury were presented in the first comprehensive study of pollutant effects on plants (von Schröder and Reuss, 1883). For example, it was reported that all signs of chronic injury occurred after 365 "single smokings" in 60 days with 1 ppm SO_2. Wislicenus (1901) found that exposure to 2 ppm SO_2 for the entire growing period caused chronic injury to young spruce trees *(Picea abies)*. Experiments were carried out in a "ventilated smoke house" probably the first fumigation chambers in which air flow rates could be regulated. Schmitz-Dumont (1896) found that a HF concentration of 3.3 ppm caused injury to pine *(Pinus silvestris)* and Norway maple *(Acer platanoides)* only after exposures of several weeks. These results, as well as those with SO_2, can only be explained through inadequate technical and analytical conditions.

O'Gara (1922), in his *Law of gas action on the plant cell* which is still often cited, derived the following formula from short-term fumigations of alfalfa *(Medicago sativa)*:

$$(c - c_R) \cdot t = k.$$

In this function, concentration (c) and exposure time (t) are directly proportional. The value c_R is the threshold concentration at which no injury occurs, even under long-term exposures. As c approaches c_R, t approaches infinity.

After extensive experimentation with SO_2, Thomas and Hill (1935) have generalized the equation so that the pollutant dose that causes a particular degree of injury can be determined. For example, under conditions of extreme plant sensitivity the following values are given:

$(c - 0.24)t = 0.94$ slight necrosis of leaves
$(c - 1.40)t = 2.10$ 50% necrosis
$(c - 2.60)t = 3.20$ 100% necrosis.

The function for the threshold concentration derived by Zahn (1963a):

$$t_R = K_1 P_1 \frac{1 + 0.5c}{c(c - c_R)}$$

represented very well the plant reactions to short-term exposures to high SO_2 concentrations which cause acute injury. Varying internal and external growth factors are taken into consideration through the resistance factors K_1 and P_1. Studies on the influence of concentration (c) and exposure time (t) have shown, however, that SO_2 does not follow the threshold law (van Haut, 1961; van Haut and Stratmann, 1960). Foliar injury increased progressively with concentration under exposure to the same products of c and t (Fig. 7).

The following example of a plant community of sunflowers, corn, field peas, and common vetch also shows that a reduction in growth correlates better with concentration than with exposure time under the same products of c and t (Guderian, 1966b). A reduction in yield of 14% at the lowest concentration level of 1 mg SO_2/m^3 air compares with a 26% reduction at 2 mg SO_2/m^3 air (Fig. 8). High concentration peaks, therefore, can have a strong negative effect on external appearance and growth and also on the composition of plant communities, especially when peaks follow one another very closely (see Sects. 2.3 and 4.1).

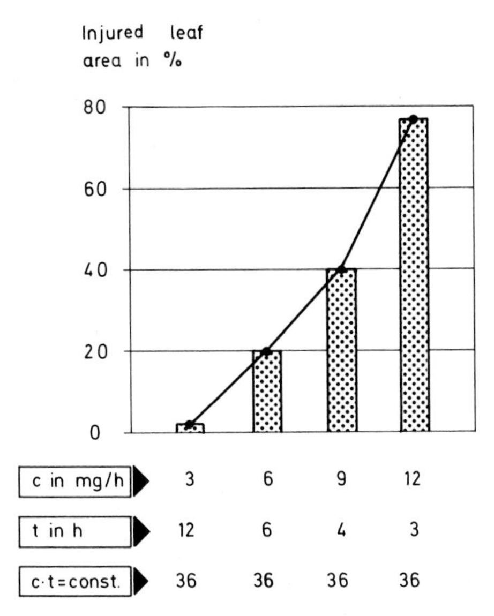

Fig. 7. Progressive increase in the degree of injury of radish with an increase in SO_2 concentration. (After van Haut, 1961)

The following exponential function, derived by Guderian et al. (1960) takes the particular risks of high SO_2 concentration into consideration:

$$t - t_R = [e^{-a(c - c_R)}] K$$

where: t = exposure time
t_R = threshold time—minimum exposure time necessary to cause injury
c = concentration
c_R = threshold concentration—minimum concentration necessary to cause injury
a = complex internal and external growth factors
K = growing time.

Products of c and t which cause injury are dependent on c in this equation and become smaller with an increase in concentration.

Effects of SO_2, which increase with the concentration, cannot be explained by an increase in sulfur accumulation in the plant. The opposite may, in fact, be true. Sulfur accumulation from uptake of SO_2 within a certain range of concentrations increases more with exposure time than with concentration, as shown in the following example (Guderian, 1970). Sulfur accumulation is understood to be the increase in sulfur above the natural sulfur content of the plant.

In experiments with pollutant amounts of 8 mg $SO_2/m^3 \cdot h$ and 16 mg $SO_2/m^3 \cdot h$ and the same products of c and t, sulfur accumulation at the lower level, with a necessarily longer exposure time, was greater than at the higher level. Experiments with longer exposure times yielded similar results (Fig. 9).

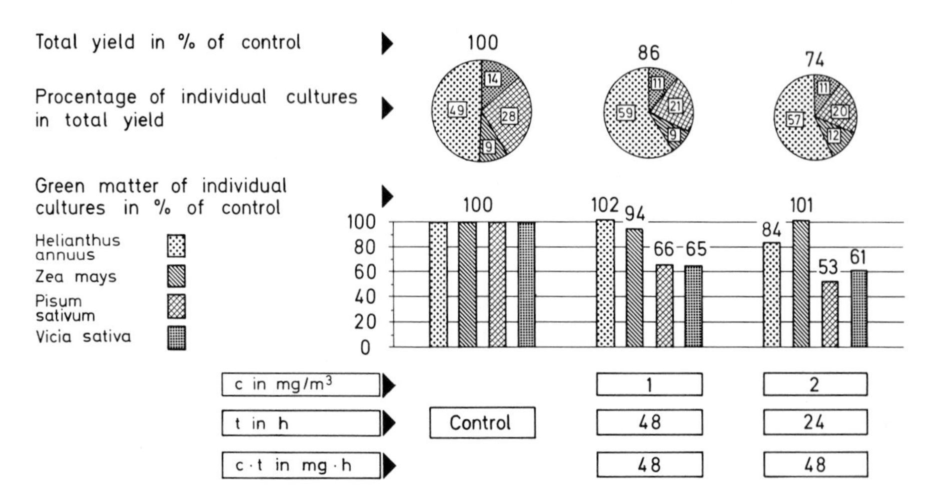

Fig. 8. Dependence of yield on concentration *(c)* and exposure time *(t)* of sulfur dioxide for a plant community consisting of sunflower, corn, field pea, and common vetch

Effects that increase with the SO_2 concentration depend not on the higher absolute pollutant uptake, but on a higher uptake rate. Per unit of time, more of the pollutant is taken up at high concentrations, so that less time is available for detoxification of the pollutant in the plant (Thomas et al., 1950a; Ziegler, 1975). If little of the pollutant is absorbed per unit of time, sulfur dioxide can be oxidized and, to a certain extent, reduced and neutralized without the occurrence of injury (Thomas et al., 1944a, b). Toxicity is greatly reduced through oxidation of the sulfite to sulfate.

The differences in absorption rates of pollutants are also manifested in the various symptoms of injury. Rapid absorption of large amounts of pollutant causes an acidification in the neighborhood of the point of entrance. This acidifi-

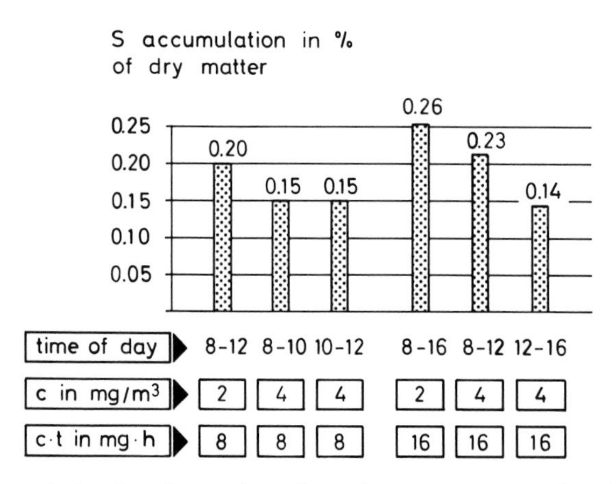

Fig. 9. Sulfur accumulation in winter wheat dependent on concentration *(c)* and exposure time *(t)* of sulfur dioxide

cation brings about a destruction of the mesophyll and formation of leaf necrosis which is usually sharply defined from the bordering tissues (van Haut and Stratmann, 1970; Jacobson and Hill, 1970). Long-term exposure to low concentrations generally results in unspecific foliar chlorosis.

Whereas injury increases with concentration within a broad range of concentrations with SO_2, contrasting results have been found in experiments with HF. Tomatoes and corn plants, which were fumigated for 513 h at 5 ppb HF, 26 h at 100 ppb HF and 4.2 h at 650 ppb HF/m^3 air showed the most severe injury and highest fluoride accumulation at the highest concentration level (Leone et al., 1956). To determine the threshold level of HF, Adams et al. (1957) carried out fumigations with 1.5 ppb, 5 ppb, and 10 ppb HF/m^3 air on 40 different plant species. The following selection gives an indication of the various different reactions observed: in Ponderosa pine, injury increased with the concentration, while just the opposite occurred with gladiolus. Larch *(Larix occidentalis)* is an example of those plants which show no clear concentration-dependent differences, either in degree of injury or in fluoride accumulation in plant material (Table 3).

The extreme difference in responses of various species to concentration and exposure time are shown with gladiolus and Ponderosa pine, which were exposed near an aluminum smelter for two weeks (Adams et al., 1956). On a site with an average concentration of 0.49 μg HF/m^3 air, gladiolus had an injury index—calculated from the relationship of necrotic leaf tip to total length of the leaf × 100—of 25.4 and on a site with 2.36 μg HF/m^3 air an index of only 32.7. Injury to pine, on the other hand, increased progressively with the concentration, as shown in the fumigation experiments cited above (Adams et al., 1957).

The differing responses are probably due to secondary translocation of the fluoride toward leaf tips and margins, which varies from species to species. In gladiolus, and other tuberous and bulb plants and some deciduous trees, fluoride is moved in the transpiration stream, and through symplasmatic and apoplasmatic transport, toward the leaf tips and margins (Jacobson et al., 1966). Injury to pine, however, is characterized by the presence of needles completely free of injury

Table 3. Total fluoride accumulation up to the appearance of the first visible foliar injury with various concentrations of hydrogen fluoride

Plant species	c in μg HF/m^3 air	t in h	$c \cdot t$ in μg HF · h	F content in dry matter in ppm
Gladiolus hortulanus	1.5	64.7	97	37
	5.0	23.8	119	46
	10.0	13.7	137	57
Pinus ponderosa	1.5	367.3	551	83
	5.0	42.6	213	72
	10.0	18.7	187	80
Larix occidentalis	1.5	71.3	107	53
	5.0	23.8	119	147
	10.0	11.8	118	106

and totally necrotic needles (Guderian et al., 1969). The common lack of tip necrosis also indicates that fluoride is probably not as strongly translocated in pine needles (see Sect. 2.5.3 and 2.5.4).

Effects of various HCl concentrations are explained with electron-microscopic studies (Masuch et al., 1973). In Experiment 1, vigorously growing six-week-old spinach plants (*Spinacia oleracea*, Matador) were exposed for 43 h within five days to HCl concentrations of 0.13 and 0.25 mg HCl/m^3 air. The higher concentration caused severe chlorosis, as well as a significant reduction of growth. The leaf taken as a sample was also chlorotic. Under the lowest concentration of 0.13 mg HCl/m^3 air, plants showed no reduction in growth and only slight chlorosis of older leaves. The sample leaf showed no differences in color intensity, in comparison to control leaves. Sample leaves, which were the same age in all replicates, had reached about two-thirds of their mature size.

In Experiment 2, 10-week-old plants were fumigated for 208 h within two weeks with 1.6 mg HCl/m^3 air. Older, fully developed leaves showed necrosis of about 1% of their area and middle leaves had slight chlorosis. The sample was taken from a chlorotic area of a fully developed leaf. The weak responses, in the form of necrosis and chlorosis, to these concentrations are due to the fact that fumigations were carried out from the end of January to the beginning of February in a greenhouse without supplementary lighting at temperatures ranging from 5° C to 14° C.

The quantitative alterations to the lamellar system of the chloroplasts are given in Table 4.

The average frequency of grana cuts in 100 μm^2 stroma area increased under influence of HCl. The average thickness of grana, that is the thickness of all thylakoids of the grana, remained constant at concentrations between 0.13 and 0.25 mg HCl/m^3 air and increased slightly at the highest concentration of 1.6 mg HCl/m^3 air.

As shown in the microphotographs in Figures 10 and 11 structural differences occurred in the organization of the thylakoid systems of fumigated and nonfumigated plants. The thylakoids of the chloroplasts in control plants in Experiment 1 had a bladderlike composition and locally stretched appearance. The interior of

Table 4. Average frequency of chloroplast components in 100 μm^2 stroma area in young (Experiment 1) and fully differentiated (Experiment 2) spinach leaves after exposure to HCl

Experiment	Grana	Lipid globules	Starch granules
Experiment 1			
Nonfumigated control	352±21	233±16	50±7
43 h · 0.13 mg HCl/m^3	441±38	340±43	42±8
43 h · 0.25 mg HCl/m^3	571±44	437±39	19±3
Experiment 2			
Nonfumigated control	450±22	472±30	24±5
208 h · 1.6 mg HCl/m^3	498±30	760±72	26±7

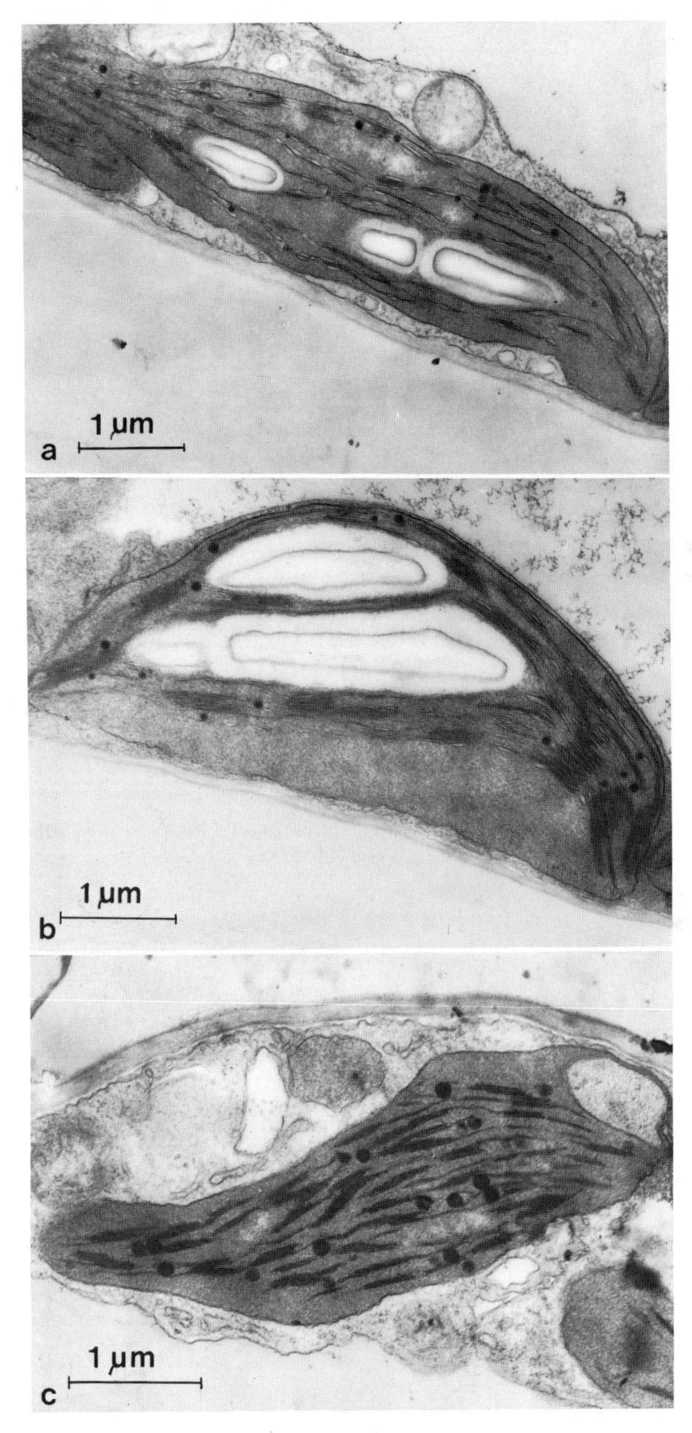

Fig. 10 a–c. Effects of various HCl concentrations on the ultrastructure of young spinach chloroplasts. (a) Control; (b) 43 h · 0.13 mg HCl; (c) 43 h · 0.25 mg HCl

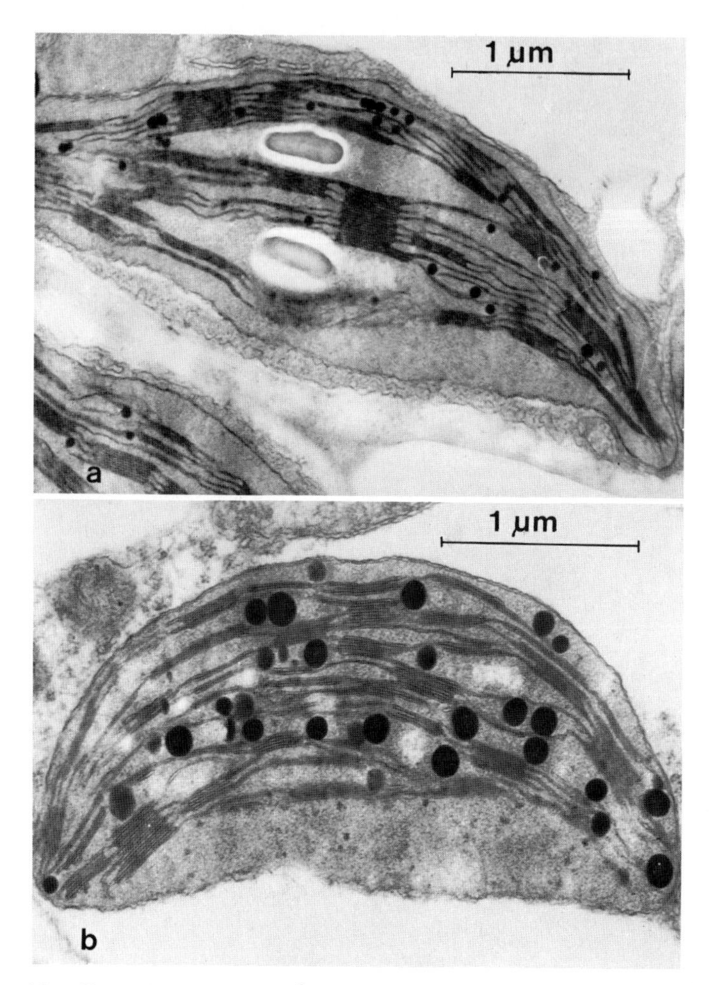

Fig. 11 a and b. Effects of 1.6 mg HCl/m^3 air on fully differentiated spinach chloroplasts. (a) Control; (b) 208 h · 1.6 mg HCl

the thylakoids of fumigated plants, especially those fumigated with 0.25 mg HCl/m^3 air, had regular light highpoints, typical for fully developed chloroplasts. The thylakoids are more clearly integrated into grana, but the grana have different diameters, so that the borders appear pointed. These low HCl concentrations, therefore, have accelerated the development of the grana.

After fumigation of fully developed chloroplasts with 1.6 mg HCl/m^3 air (Experiment 2), the light highpoints of the cisterna doubled in size in all grana thylakoids (Fig. 11), causing a greater average thickness of grana in comparison to control plants. The imbibing effect of chloride ions, therefore, can even be seen in the ultrastructure of plant organs.

Effects of HCl are particularly clear on the osmiophilic lipid globules. They increase in size and number with an increase in concentration, as shown in Table 4 and Figures 12 and 13.

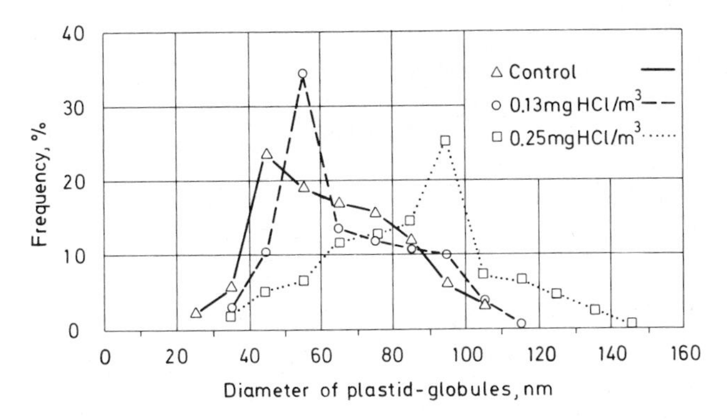

Fig. 12. Effect of low HCl concentrations on the size-distribution of plastid-globules in spinach chloroplasts

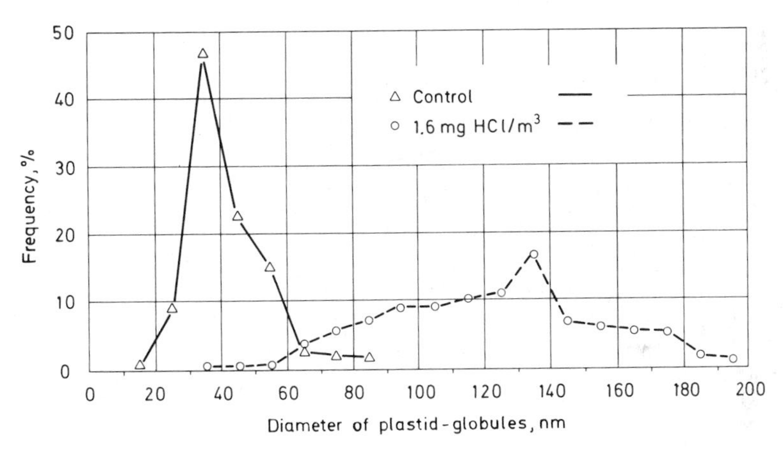

Fig. 13. Effects of 1.6 mg HCl/m^3 air on the size-distribution of plastid-globules in spinach chloroplasts

In summary, it can be said that especially chronic exposure to HCl causes changes in the ultrastructure of chloroplasts, which are similar to changes that occur during normal degenerative processes. Different HCl concentrations, therefore, accelerate the aging processes to a certain degree.

Under chronic exposure to SO_2, other structural alterations in the chloroplasts occur. The stroma plasma becomes granular with plasma agglomerations. The osmiophilic lipid globules increase neither in size nor in number. On the borders of the grana thylakoids, however, a greater number of osmiophilic granules were observed. Osmiophilic granules also occur between the stroma thylakoids, and sometimes there is contact between the two. Swelling of the thylakoids and an increase in volume of the chloroplasts, observed by Fisher et al. (1973) after short exposures to SO_2 concentrations which cause acute injury, were not observed after long-term exposure to low concentrations.

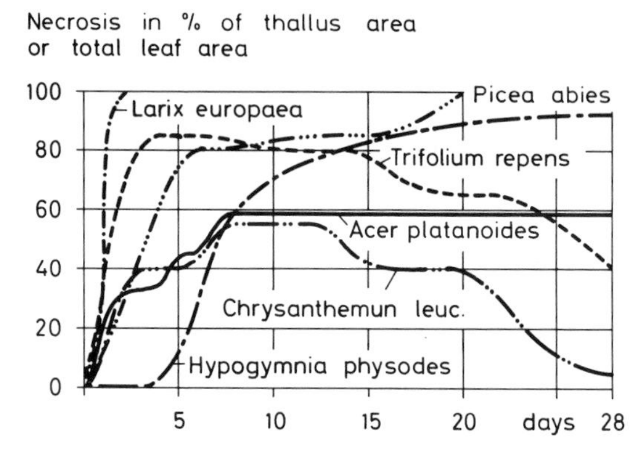

Fig. 14. Injury to higher plants and to the foliose lichen *Hypogymnia physodes* at a concentration of 1.9 mg SO_2/m^3 air

In experiments in which spinach was fumigated with a combination of SO_2 and HCl, structural alterations occurred that were typical for the single components. Intensity of the effects was also increased (see Sect. 2.4). Electron-microscopic studies, therefore provide a basis for differentiating between effects of SO_2 and HCl.

The threshold curves already described show only the beginning of injury, usually to the leaves. For further effects to the plant, type and extent of injury development above the threshold are the determining factors. As can be seen in Figure 14, three different types of response occur in the development of injury above the threshold time: on larch, spruce, and the foliose lichen *Hypogymnia physodes*, necrosis increases with increasing exposure times up to total injury, whereas after a specific exposure time no further external alterations in leaves of Norway maple occur. Because of the formation of new foliage during fumigation which is less affected, total injury of white clover and chrysanthemum decreases after a specific exposure time. In the physiologic causes of these different reactions, regenerative capacity of the plant is quite important. Since gas exchange is reduced under influence of pollutants, the newly formed leaves do not absorb enough pollutant to cause injury.

Results reported here, and similar results from experiments with ozone (Heck et al., 1966; Ting and Dugger, 1968) and with irradiated NO_2-propylene mixtures (Heck, 1964) indicate the difficulties involved in establishing injury curves above the threshold limits. When it is considered that—contrary to other opinions (Brisley and Jones, 1950; Thomas, 1961)—it is difficult, when not impossible (Guderian and Stratmann, 1968), to obtain reliable information on the decrease in useful value of a particular plant species under practical conditions because of the extent of foliar injury, it is understandable why formulas derived from controlled fumigation experiments are used less for a direct characterization than for an indirect evaluation of pollutant conditions in the field (see Sects. 2.3 and 4.1). It can be seen that mathematical formulation of experimental results becomes more difficult as the complexity of the test object increases (Cernusca, 1971) as is the case under constantly changing pollutant and climatic conditions in the field.

2.3 Significance of Continuous and Intermittent Pollutant Action

Concentration and duration of action of atmospheric pollutants change constantly and are dependant on the type of pollutant complex. Near single sources, fluctuations in concentration are particulary great (Fig. 15), and very high concentration peaks, as well as very low concentrations or even pollutant-free periods can be observed. With an increase in distance from the source, the exposure time, that is the duration of action of a pollutant, is reduced compared to the total time.

Heavily industrialized areas, however, are usually characterized by a typical pollutant complex and pollutant loading occurs independently of wind direction. In such situations, the concentration curve is usually more uniform when the source density is high, that is a large number of sources of a particular pollutant in a single area (Dreyhaupt, 1970). Extreme fluctuations in concentration can also occur, however, especially in connection with certain meteorologic parameters. Relatively constant concentrations can be monitored during inversion situations (Schwarz, 1961).

The concentration/time relationships described in Section 2.2 are particularly valid for continuous action of pollutants. For the analysis of intermittent pollutant action, it is important to determine if, and which, changes occur in these relationships when action of a certain gas is interrupted. Theoretically, a positive effect for vegetation is to be expected when exposure is interrupted by pollutant-free periods. For example, extremely toxic SO_2, because of its neutralization and redox potentials, is oxidized to the less toxic sulfate. If accumulated SO_2 exceeds this capacity, acute injury occurs, but when both processes are equalized, no necrosis occurs. This neutralization happens relatively quickly. Thomas et al. (1950a) have shown that in sunlight, 1–2 h after fumigation, no SO_2 could be detected in leaf mesophyll.

Figure 16 shows the different reductions in CO_2 uptake as a result of continuous and intermittent exposure to SO_2 (Guderian, 1970). In this experiment winter barley *(Hordeum vulgare)* was exposed to 4.1 mg SO_2/m^3 air one time for 2 h (1×2), two times for 1 h (2×1), four times for $^1/_2$ h ($4 \times {}^1/_2$), and eight times for $^1/_4$ h ($8 \times {}^1/_4$). Concentration and duration of exposure, and, therefore, pollutant load, were constant, but the pollutant-free periods were of different lengths. During the exposures 1×2 and 2×1, the CO_2 uptake fell to 20% of the control. Photosynthesis during the $4 \times {}^1/_2$ and $8 \times {}^1/_4$ exposures was reduced only about 50%. On the following day, without additional fumigation, plants under the $4 \times {}^1/_2$ and $8 \times {}^1/_4$ regimes had recovered almost normal photosynthesis, whereas plants under the 1×2 and 2×1 regimes still showed severely reduced photosynthesis, although gradual recovery was apparent.

The recovery of photosynthesis in crimson clover *(Trifolium incarnatum)* half an hour after fumigation with a constant SO_2 concentration for various exposure times is shown in Figure 17. The plants exposed for $^1/_2$ h only show a clear recovery increase in CO_2 uptake, whereas those exposed for 1 h show a slight recovery in photosynthesis, and plants exposed for 1 and $^1/_2$ h did not recover at all.

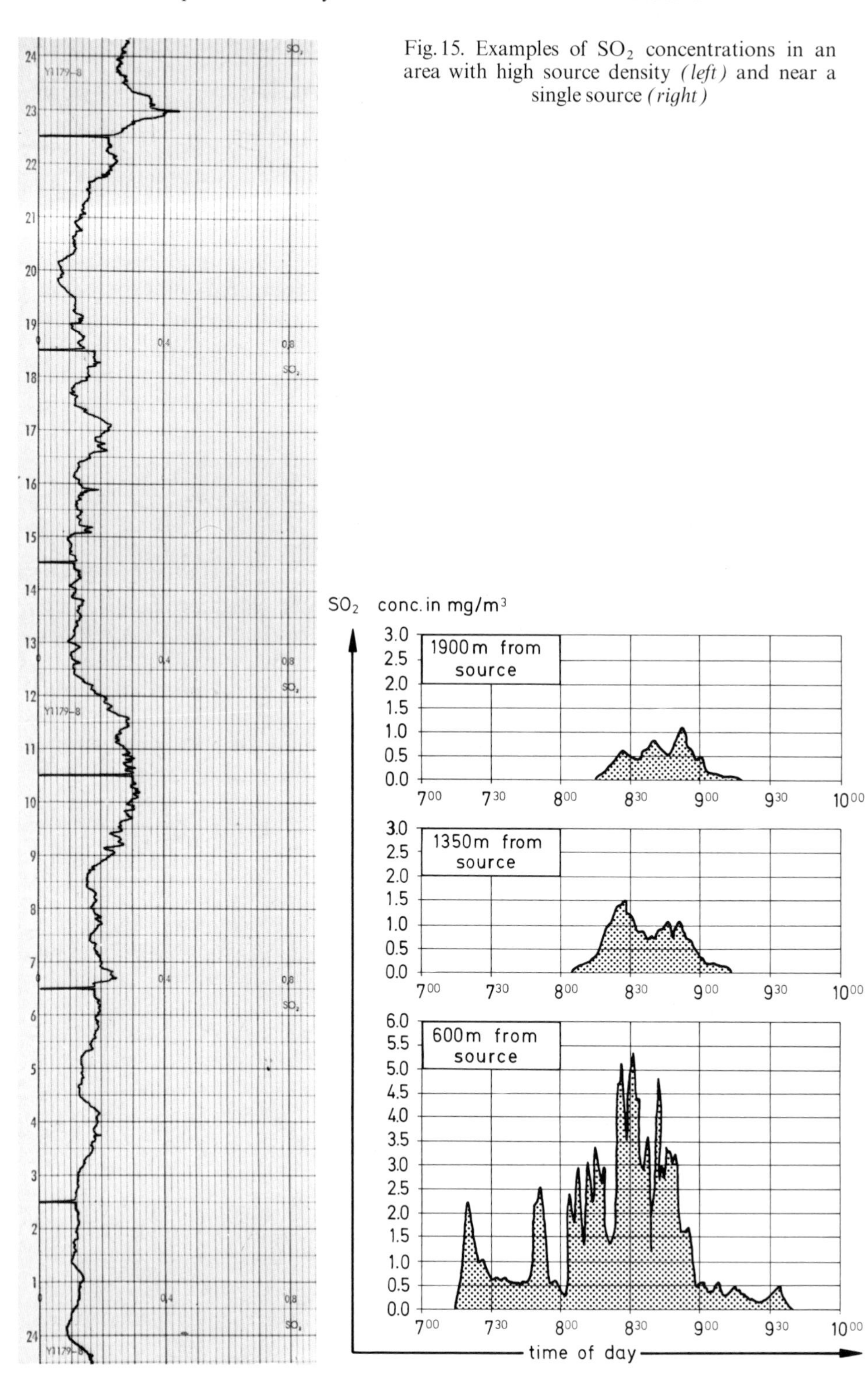

Fig. 15. Examples of SO_2 concentrations in an area with high source density *(left)* and near a single source *(right)*

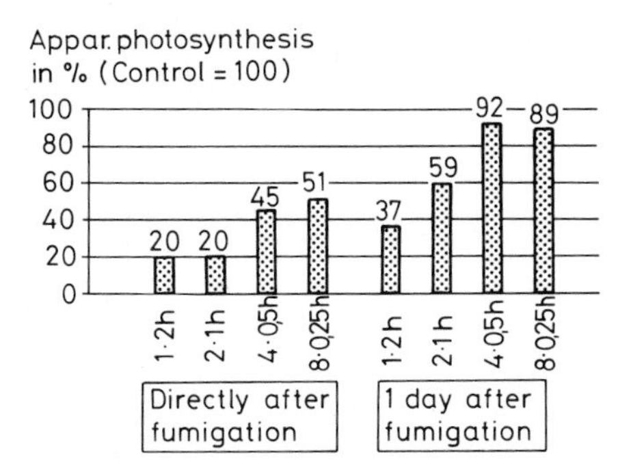

Fig. 16. Apparent photosynthesis of winter barley dependent on frequency and length of exposure to SO_2 at a constant concentration

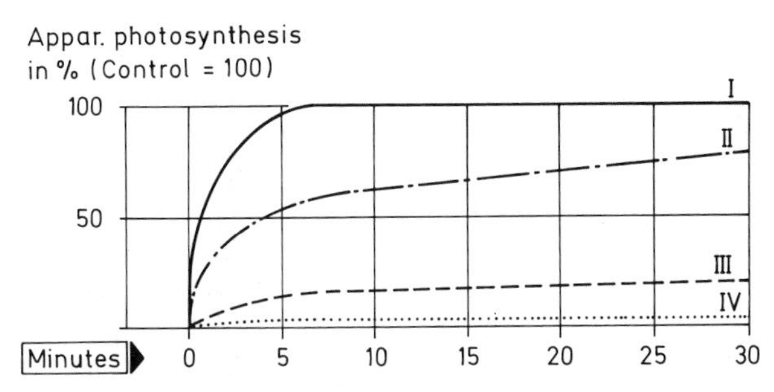

Fig. 17. Apparent photosynthesis of crimson clover after exposure to 3.8 mg SO_2/m^3 air for different times. *I* Control; *II* 0.5 h·3.8 mg SO_2; *III* 1 h·3.8 mg SO_2; *IV* 1.5 h·3.8 mg SO_2

Experiments demonstrate that recovery can occur during pollutant-free periods. However, at a particular concentration, such a recovery is only possible when a definite exposure duration has not been exceeded. In agreement with Zahn (1961, 1963a), who investigated the influence of intermittent fumigations on leaf injury, it has also been shown that pollutant-free periods must reach a certain minimum to allow recovery.

The less severe effects observed in experiments with intermittent exposures to constant SO_2 concentrations are not due to reduced pollutant uptake. In general, S-accumulation increases with an increase, up to a certain frequency, in pollutant impulses (Guderian, 1970). In fact, MacLean et al. (1968) have shown that red clover *(T. pratense)* and timothy *(Phleum pratense)*, when exposed intermittently to HF, accumulated twice as much F as plants exposed continuously. In both cases, the extent of injury was the same.

The reduced effects of intermittent exposures to pollutants may also be due to a reduction of pollutant content in the plant during pollutant-free periods. Especially fluoride, which is present in the plant in soluble form and maintains the chemical properties of free inorganic F-compounds (Jacobson et al., 1966), can be washed and excreted out of the plant tissues. During a pollution-free week, the F-content of the gladiolus variety Snow Princess was reduced by 40–50% (Hitch-

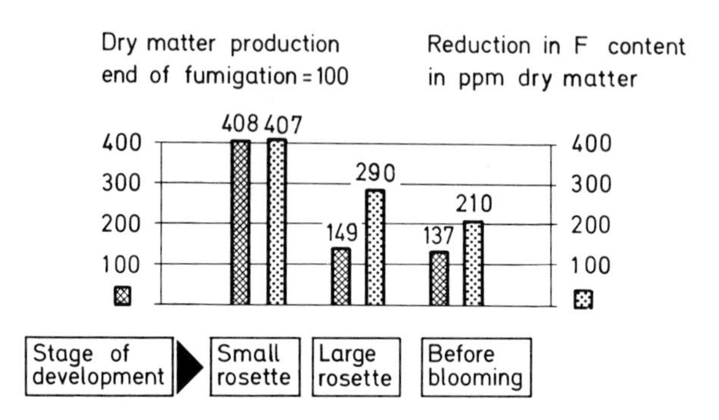

Fig. 18. Decrease in fluoride content of summer rape in various stages of development after exposure to HF compared with dry matter production

cock et al., 1962) and the F-content of spruce *(P. abies)* was reduced from 240 ppm to 40 ppm in dry matter in five months (Knabe, 1970a). Extreme variations in Cl content also occur due to washing effects.

Figure 18 shows the influence of plant productivity during pollutant-free periods on pollutant content of plants (Guderian et al., 1969).

The fluoride content of rape *(Brassica napus* var. *oleifera)* in various stages of development was most strongly reduced when productivity during pollutant-free periods was greatest. The F-content of the older, more slowly growing plants was least affected and dropped from 315 ppm to 105 ppm in dry matter. In younger plants, with more intensive growth, the F content was reduced from 469 ppm to 62 ppm in dry matter. Guderian (1970) has also shown that dilution effects occur after exposure to SO_2 through production of plant material low in natural S content.

2.4 Effects of Gaseous Air Pollutants in Combination

When evaluating effects of air pollutants, one must differentiate between the action of single pollutants occurring alone, alternately, or in combination. Although plant responses to single pollutants have been extensively investigated over several decades, the effects of pollutants in combination present a completely new field of study. Such effects include plant responses to several pollutants acting at the same time or alternately. In heavily industrialized areas, a mixture of various pollutants always occurs, whereas in less industrialized areas, usually as a result of changes in wind direction, single pollutants may occur one after the other. Theoretically, it can be assumed that under such pollutant conditions, the effects of the single components are either reduced or increased, or that additive effects occur.

Experiments with oats (*Avena sativa*, Flämings gold) red clover (*T. pratense*, Red Head), and spinach (*S. oleracea*, Matador) have shown that strong synergistic reactions occur when fumigated with SO_2 and HCl at the same time. Table 5 presents results of experiments carried out in the fumigation chambers described in Section 1.1.3. The stimulation of effects can be clearly seen.

Table 5. Leaf injury and apparent photosynthesis of spinach exposed to SO_2 and HCl alone and in combination

Exposure time in h	Time of day	Necrosis in % of total area				Apparent photosynthesis in ppm		
		SO_2	HCl	SO_2+HCl	Control	SO_2	HCl	SO_2+HCl
19	9^{00}	—	—	—	34	35	34	31
23	13^{00}	—	—	—[a]	33	34	32	29
42	8^{00}	—[a]	—[a]	8	36	36	31	26
53	19^{00}	0.5	1	25	28	25	23	14

[a] Appearance of first necrotic lesions.

Spinach plants, in the eight-leaf stage, were fumigated for 53 h with 0.9 mg SO_2 and/or 1.35 mg HCl/m^3 air. Air exchange in the chambers was 50 times/h. One chamber had only SO_2, a second only HCl, and a third both SO_2 and HCl in the same concentrations (0.9 mg SO_2, 1.35 mg HCl). The fourth chamber served as the control. Duration of the light period was 12 h. At the beginning of the light period on the second day, the plants exposed to both components showed the first significant decline in CO_2 uptake, which became greater throughout the entire experiment. At the end of 53 h, photosynthesis of these plants dropped 14 ppm to about 50% of the control plants. Photosynthesis of plants under SO_2 alone was reduced 3 ppm and 5 ppm under HCl alone.

The degree of leaf injury of plants fumigated with both components is even more significant. Necrosis of 0.5 and 1.0% for plants under SO_2 and HCl compares with 25% for plants under SO_2 and HCl together.

An increase in effects has also been shown for combinations of other pollutants. Menser and Heggestad (1966) showed that injury to tobacco occurred after fumigation with 0.03 ppm ozone and 0.30 ppm SO_2 for 2 or 4 h. No injury occurred to plants exposed to SO_2 and ozone alone in these concentrations. Similar effects have also been found by Menser and Hodges (1970), Grosso et al. (1971), and Hodges et al. (1971). MacDowall and Cole (1971) have reported an increase in effects with a combination of ozone and SO_2 that was below the threshold level of SO_2. Applegate and Durrant (1969), working with peanuts *(Arachis hypogaea)*, found that the synergism between ozone and SO_2 caused other symptoms of injury than the single pollutants alone. Tingey et al. (1973) also found synergistic effects of these two pollutants in experiments with tobacco, radish *(Raphanus sativus)*, and alfalfa *(Medicago sativa)*.

Exposure of various species to sulfur dioxide at concentrations less than 0.5 ppm or nitrogen dioxide at concentrations less than 2.00 ppm for 4 h caused no injury. A combination of these pollutants caused foliar injury at concentrations between 0.05 and 0.25 ppm (Tingey et al., 1971). Matsushima and Taylor (cited in Matsushima and Brewer, 1972) also observed a strong synergism between SO_2 and NO_2. An increase in effects was also shown through changes in the activity of certain enzymes in pea seedlings *(Pisum sativum)* at concentrations of 0.2–2 ppm SO_2 and 0.1–1 ppm NO_2 (Horsmann and Wellburn, 1975).

Combined pollutants cause not only synergistic, but also additive effects. The degree of injury brought about by a combination of pollutants represents the sum of effects which would be caused by single pollutants, as shown in experiments with orange varieties (*Citrus sinensis* and *C. unshu*) fumigated with SO_2 and HF together (Matsushima and Brewer, 1972). Some species, such as white cabbage (*Brassica oleracea capitata*), cauliflower (*B. oleracea botrytis*), and tomato, exhibit additive effects from the combined action of SO_2 and ozone (Tingey et al., 1973), while other species, as mentioned above, show synergistic effects.

A reduction of effects very seldom occurs. According to Haagen-Smit et al. (1952), SO_2 reduced the phytotoxicity of a mixture of gasoline fumes and ozone. Tingey et al. (1973) have also found reduced effects with mixtures of SO_2 and ozone, but additive and synergistic effects predominate.

The extent of effects from alternately occurring pollutants seems to depend on the order in which the pollutants occur. Exposure to NO_2 and then to SO_2 caused no change in the effects of SO_2, but a strong increase in effects of SO_2 was observed when plants were exposed first to SO_2 and then to NO_2 (Matsushima and Taylor, cited in Matsushima and Brewer, 1972).

The relatively few studies on effects of pollutants in combination allow no quantitative statements on the hazards to vegetation exposed to such pollutant complexes. In general, however, the risk of injury to vegetation is greater when several pollutants occur together.

2.5 Significance of External and Internal Growth Factors

External growth factors are those determined through edaphic and climatic conditions. Internal growth factors are expressed in the specific degree of resistance of the plant. Resistance varies with environmental conditions and stage of development of the plant.

The fundamental importance of these factors on the susceptibility of vegetation to atmospheric pollution was recognized quite early (von Schröder and Reuss, 1883; Wislicenus, 1898). However, only little information could be obtained for clarification of these complex problems, so long as observations and inquiries were confined to effects on vegetation in polluted areas. Only after development of chemical–physical means for air analyses and experimental apparatus for fumigation, with the possibility of controlling pollutant concentrations as well as other growth factors, did it become possible to study the influence of single factors on susceptibility of plants. Fumigation experiments in the field and in the laboratory forced investigation of these questions, as it is particularly important to determine to what extent knowledge gained from experiments under artificial conditions can be applied to ambient conditions. The following summary and evaluation of results taken from the literature and from the author's experiments should aid in the clarification of some of these problems, especially in connection with dose–response relationships.

2.5.1 Influence of Climatic Factors

Climatic factors before, during, and after exposure determine, to a large degree, the type and extent of plant responses to a given pollutant. Certain climatic conditions on the days preceding exposure increase plant susceptibility, others reduce it. Climatic conditions during exposure are most important and have a major influence on uptake and effect of the pollutant. After exposure, climatic factors have a lesser influence on plant responses. From the complex of climatic factors, single physically definable parameters, such as light, temperature, and humidity, are discussed.

2.5.1.1 Radiation

It is the general opinion that gaseous air pollutants enter the leaf mesophyll primarily by diffusion through the stomata (Kisser, 1966; Thomas and Alther 1966; Ting and Dugger, 1968). All external factors that regulate stomatal movement will influence uptake of atmospheric pollutants by plants. Radiation is one of the most important of these factors as,when all other factors are held constant, stomatal reactions to light can be interpreted as photonasty (Mohr, 1969).

Because the stomata of most plants are closed in darkness, it was generally thought that pollutant action during the night would have little or no effect (Lotfield, 1921; Zimmermann, 1950). Van Haut (1961) and Zahn (1963b), however, were able to disprove this assumption in experiments with SO_2. Although action of pollutants during the night has a less severe effect than during daylight, both effects must be considered together, as exposure during the night can intensify the effects of daylight exposures.

In experiments on uptake of SO_2 by plants during light and dark periods, it was shown that S accumulation at night can be about one third of that during daylight (Guderian, 1970). These high rates of accumulation also explain the response of plants to exposures during darkness.

Table 6. Cl accumulation and degree of injury to *Lolium multiflorum*, *Trifolium pratense* and *Lycopersicum esculentum* exposed to HCl during light and dark periods[a]

Plant species	Control Cl content in % dry matter	Light period		Dark period	
		Degree of injury	Cl content in % dry matter	Degree of injury	Cl content in % dry matter
Lolium multiflorum	1.21±0.08	—	2.38±0.07	—	1.58±0.03
Trifolium pratense	0.70±0.10	0.8±0.4	1.67±0.14	0.3±0.2	0.86±0.08
Lycopersicum esculentum	0.87±0.06	—	2.25±0.16	—	1.93±0.10

[a] Average of five samples.

Similar relations between light and dark periods occur in the accumulation of chloride, as shown inTable 6.

In this experiment, carried out in controlled-environment chambers, the plants (Welsch rye grass, Tiara; Red clover tetraploid, Red Head; tomato, Sterlingcross F1) were exposed to 1.2 mg HCl/m^3 air for 48 h for one 12-h period on each of 4 consecutive days with the light period from 7.00–19.00 and the dark period from 19.00–7.00.

The Cl accumulation, that is the difference in Cl content between fumigated and nonfumigated plants, varied greatly, depending on the species. Although accumulation in red clover during darkness, with 0.16% Cl in dry matter, reached only about one sixth of the accumulation in daylight (0.97% Cl) the relation in rye grass, with values of 0.37% and 1.17% Cl, respectively was much closer. It is apparent that the stomata, assumed to be closed in darkness (Meidner and Mansfield, 1968), still allow considerable gas exchange.

Of particular interest is the pattern of injury on red clover from exposure to HCl in darkness. It must still be determined through further investigation if Cl, like NO_2 (van Haut and Stratmann, 1967; Taylor, 1968), has a greater phytotoxicity in darkness than in daylight. As expected, the difference between Cl uptake in darkness and daylight in tomato was small. Stomata of plants of the Solanaceae normally remain open at night and it is clear from this example that pollutant uptake is strongly dependent on the size of the stomatal opening.

Adams et al. (1957) have studied fluoride accumulation and extent of foliar injury during fumigations with HF in darkness and in daylight. The first foliar injury from exposure in darkness occurred on 40 different plant species or varieties with a product of concentration and exposure time that was only 10% higher than that necessary to cause injury in daylight. The F accumulation varied greatly from species to species, but the average F accumulation necessary to cause injury was only 53% of that required in daylight exposures.

The exceptional effectiveness of fluoride taken up in darkness probably resulted from the unusually long dark period. Plants were exposed in darkened chambers at the same time as plants in daylight. Follow-up experiments are necessary to determine if plants show increased sensitivity in darkness. In experiments on alfalfa *(M. sativa)* by Benedict et al. (1965), fluoride uptake in darkness reached 40% of daylight values, although stomata were closed, as shown by porometer measurements. Action of fluoride during the night presents a special hazard for vegetation, as uptake is greater than that of SO_2 and HCl.

Experiments on the influence of light intensity on plant susceptibility have shown that a positive correlation exists between foliar injury and increasing light intensity up to 38000 lux during exposure. This has been found to be true with SO_2 (Setterstrom and Zimmermann, 1939) and with various components of photochemical smog (Juhren et al., 1957; Heck et al., 1965; Heck, 1966). For some species, sensitivity to SO_2 (Zimmermann, 1950) and to ozone (Heck, 1966; Heggestad and Heck, 1971) increases up to full sunlight intensity.

Little information is available concerning the effects on the interaction of light quality with phytotoxicants on plant sensitivity. Research with other gases and information on the connection between light quality and stomatal movement suggest that light quality does affect sensitivity. Bean plants proved to be more

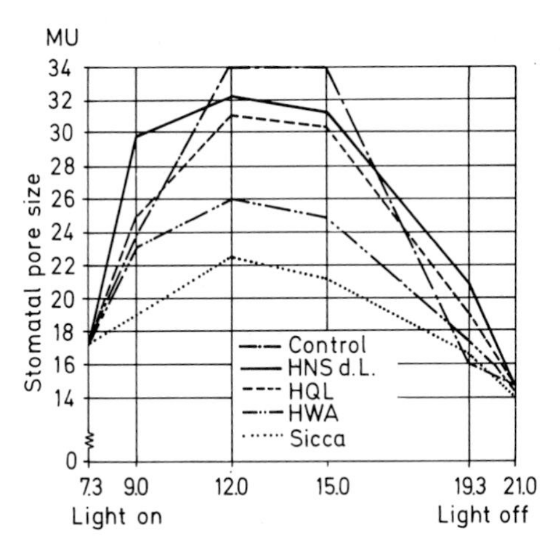

Fig. 19. Daily variations in the size of the stomatal pores of lettuce grown under various light conditions (in units of measure = *ME*). (After Vogel, 1960)

sensitive to the photochemical component PAN (peroxyacetylnitrate) under wavelengths of 420 and 480 nm than under 640 nm (Dugger et al., 1963). In further research, Dugger and Ting (1968) showed that PAN sensitivity is controlled by a light-mediated system in the 660 nm and 700 nm range and that plants were more sensitive under 660 nm than under 700 nm.

The connection between pollutant uptake and stomatal movement has already been mentioned. In experiments on lettuce under various light fields, Vogel (1960) has shown that the spectral composition of electromagnetic radiation has a strong influence on the degree of opening of stomata and rhythm of stomatal movement (Fig. 19).

In light high in long-wave reds (Sicca, HWA), the stomata remained more tightly closed during the entire day than those of the control in normal daylight. Light (HQL, HNS, dL) with a spectral composition similar to the "basic-curve" (Ruge, 1958) with a first maximum between 410 and 430 nm, a second maximum at 660 nm, and a minimum between 450 and 490 nm, caused an earlier opening and later closing of stomata in comparison to daylight. In their research with different wave lengths, Mansfield and Meidner (1965) found an increase in pore size from the red to the blue wave lengths. Kuiper (1964) and Meidner (1968) have shown that the action spectrum of the stomata, with a maximum pore size at 432 and 675 nm, is similar to the absorption spectrum of the chlorophylls and the action spectrum of photophosphorylation.

In summary it can be said: the dynamic equilibrium between the light-dependent photoactive opening and hydroactive, but also passive, closing determines the stomatal pore size. Quality and intensity of radiation play a decisive role in this process and, therefore, also in pollutant uptake through leaves. Depending on the type of artificial illumination, plants react with different sensitivities to a given air pollutant.

The degree of stomatal opening gives an indication as to the degree of risk of injury by a particular pollutant. It must not be overlooked, however, that there is no linear relationship between pore size and permeability of leaves to gases (Ting and Dugger, 1968). Diffusion of a gas through stomata can be described, in a simplified form, by the following Eq. (1):

$$Q=\frac{D\cdot\Delta p}{R_L+R_B} \tag{1}$$

where: Q = gas-diffusion rate, i.e., μg/cm²/sec
D = diffusion coefficient of the gas
Δp = partial pressure difference between the atmosphere and the inside of the leaf in μg/cm³
R_L = leaf or stomatal resistance in sec/cm
R_B = boundary layer resistance, independent of stomatal characteristics.

The R_L value can be measured directly with a resistance hygrometer (Ting et al., 1967) or computed, when the number of stomata, their distance from one another, and pore size are known (Lee and Gates, 1964). The only variable in the R_L value of a leaf is the stomatal aperture. The R_L value varies inversely, but not linearly, with the pore radius, as shown in Eq. (2).

$$R_L=\frac{1}{n}\left(\frac{l}{\pi r^2}+\frac{1}{r}\right) \tag{2}$$

where: l = pore length
r = pore radius
n = number of pores/unit area.

It follows that changes in the stomatal aperture do not necessarily lead to corresponding changes in gas uptake. If the pore radius, with mostly closed stomata, is small, the R_L value is large; the opposite is true when the stomata are wide open.

Knowledge of stomatal movement alone allows only a rough estimation of pollutant uptake, as external parameters, such as air pressure and wind speed have a strong influence on factors D and R_B in Eq. (1) (Meidner and Mansfield, 1968). Besides that, the determination of representative stomatal responses for an entire leaf is quite difficult and requires extensive experimental investment. Domes (1971) has shown that the guard cells on the upper and lower leaf surfaces react differently to external parameters. When possible, reliable methods of chemical analysis are used, therefore, for the determination of pollutants taken up through the leaf and incorporated in plant tissues. No predictions as to hazards to vegetation are possible with these methods, however.

Along with effects on stomatal function, radiation during leaf development also influences size and number of developing stomata. It can be said that light conditions during plant development determine the predisposition of plants to atmospheric pollutants. This factor has a greater importance the longer plants are held under artificial light.

Table 7. Cl accumulation in *Lolium multiflorum* under constant climate conditions

Time of day	Cl content in % of dry matter[a]
5– 7	1.29±0.26
7– 9	1.27±0.22
9–11	1.38±0.17
11–13	1.32±0.08
13–15	1.32±0.18
15–17	1.22±0.16
17–19	1.29±0.08

[a] Average of five samples.

In addition to the effects of light quality on stomata, it is assumed that light composition has a strong influence on the sensitivity of vegetation to air pollutants, especially when the effects of radiation on various radiation-dependent physiologic processes are considered. It is known from observations in polluted areas and in fumigation chambers that high light intensity, not only during, but also after exposure, can intensify the reactions of plants to air pollutants (Stoklasa, 1923; Haselhoff et al., 1932; van Haut and Stratmann, 1970). For example full sunlight, following an exposure of soybeans (*Glycine max* L. Merr) to HF accentuated injury, while shadow reduced it (Wiebe and Poovaiah, 1973).

The daily rhythms in pollutant accumulation and in plant sensitivity are essentially light-dependent processes. In fumigation experiments in the field Katz (1949), Thomas (1951), and Zahn (1963b) have shown that plants respond differently to the same SO_2 concentration at various times of day. Van Haut (1961) has observed a daily rhythm even under constant light intensity, temperature and humidity conditions. HCl accumulation, however, varied only sightly under constant growth chamber conditions, as shown in Table 7.

Welsh ryegrass (Lema), after the second cutting, was exposed on four days during the light period to 1.0 mg HCl/m^3 air under constant climatic conditions (temperature 21.6° C, rel. humidity 49%, light intensity 22000 lux).

Thomas and Hendriks (1956), in agreement with Katz (1949), relate SO_2 sensitivity to the degree of opening of stomata. All factors which induce opening of stomata, such as light, optimal water supply, high relative humidity, and optimal temperature, increase the effectiveness of a given pollutant. Comparative experiments on diurnal rhythm, however, have shown that differences in leaf injury cannot always be explained by differences in SO_2 uptake (Fig. 20). In an experiment, carried out in plastic chambers in the field, resistance of oats (*Avena sativa*, Flämingsgold) was clearly increased during the day when sulfur accumulation and leaf injury are compared.

Increase in resistance during the day appears to be connected with sugar content of the leaves (Guderian, 1970), as shown in experiments in which bean plants *(Phaseolus vulgaris nanus)* in different glucose solutions, were exposed to SO_2 (Fig. 21). Dugger et al. (1962, 1963) also found a clear correlation between sugar content of leaves and resistance to ozone.

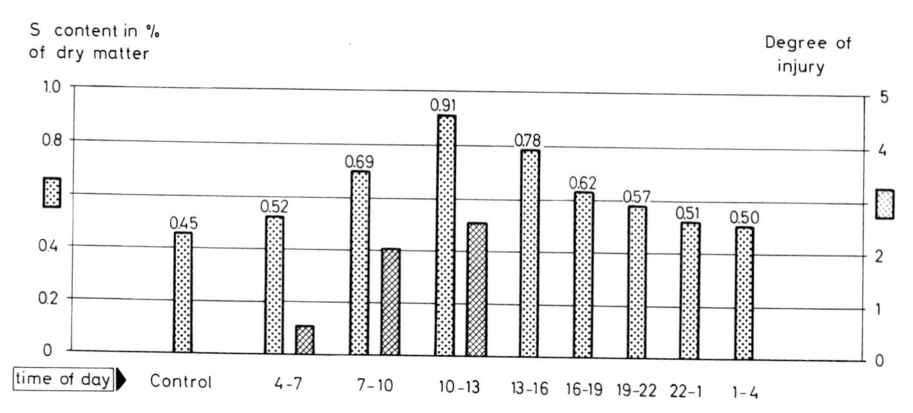

Fig. 20. Sulfur accumulation and degree of injury to oats exposed to SO_2 at various times of day

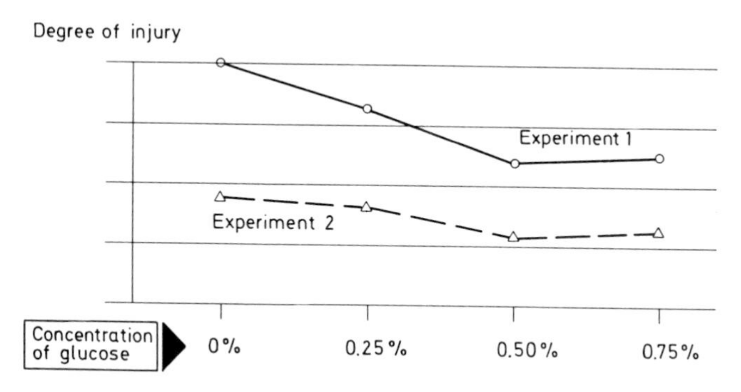

Fig. 21. Influence of the concentration of various glucose solutions on the sensitivity of field bean leaves to sulfur dioxide

This connection between carbohydrate content of leaves and plant sensitivity is demonstrated with the following experiment on tomato (*Lycopersicum esculentum*, Sterlingcross F1) exposed to HCl (Table 8).

After exposure to 2.4 mg HCl/m^3 air for 4 h, leaflets of plants held in darkness (dark plants) for 36 h before exposure showed 25% necrosis, while leaflets with a normal light/dark rhythm (light plants) were not injured. These differences come from variations in sensitivity. The Cl accumulation of the light plants, with 0.42% Cl in dry matter, was one third higher than in the dark plants. Exposure to pollutants during morning hours, therefore, presents a particular hazard for vegetation. This is especially true under practical conditions, as under daily periodic variations in meteorologic exchange processes the pollutant concentrations reach a distinct maximum at this time (Kuelske and Prinz, 1968).

2.5.1.2 *Temperature*

Various growth and development processes, especially the dark reactions in photosynthesis, represent temperature-dependant chemical reactions. This ex-

Table 8. Leaf injury and Cl accumulation in tomato after exposure to HCl in darkness for various lengths of time

Treatment	Cl content in % dry matter[a]	Cl accumulation in % dry matter	Necrotic area in % total leaf area[b]
Control			
Light plants	0.92 ± 0.07		
Dark plants	0.97 ± 0.06		
4 h · 2.4 mg HCl/m³			
Light plants	1.34 ± 0.06	0.42	—
Dark plants	1.29 ± 0.08	0.32	$24 \pm 5\%$

[a] Average of four samples.
[b] Average of eight plants.

plains the fact that, up to about 30° C, photosynthesis of many plant species has a Q_{10} of 2 when supply of CO_2 and light is high (Mohr, 1969). In extensive investigations, it could be shown that the relationship between CO_2 assimilation and temperature has the form of an optimum curve (Stalfelt, 1960). The intensity of photosynthesis influences the size of the stomatal pore through the partial pressure of CO_2 in the intercellular spaces (Heath and Russell, 1954; Jones and Mansfield, 1970). The size of the stomatal pore is also directly affected by temperature (Lange, 1975) and an influence on pollutant uptake is, therefore, to be expected. Temperature can also have a direct influence on the effects of pollutants absorbed by mesophyll cells.

Available information does not allow an intensive evaluation of the actual influence of temperature on the effectiveness of given concentrations of SO_2, HCl, or HF. This is true for effects of temperature before, during, and after exposure to pollutants. Through an analysis of experiments with photochemical oxidants on the effects of temperature conditions during plant cultivation (Kendrick et al., 1953; Juhren et al., 1957) it can be said that low, as well as high, temperatures one or several days before exposure reduce plant sensitivity, whereas normal temperatures increase sensitivity. Four tobacco varieties, which were held for two weeks before exposure to ozone under a daytime temperature of 20° C and a night-time temperature of 15° C, were more resistant than plants raised under temperatures of 25° C and 20° C. Because of the connection between sensitivity and carbohydrate content of leaves (see Sect. 2.5.1.1), increased sensitivity due to high temperatures could occur as a result of higher respiration rates at these temperatures. Especially during long cultivation under various temperature conditions, the influence of different stages of development cannot always be separated from direct effects of temperature on plant sensitivity.

Under greenhouse conditions and especially in the field, it is almost impossible to differentiate between the effects of light intensity and temperature (Zahn, 1963b) as there is a positive correlation between the two factors (Juhren et al., 1957). It is generally assumed that plant sensitivity increases with temperature over a wide range, from about 4° C to 35° C. In growth chamber experiments under constant light conditions, however, Heck et al. (1965) and Cantwell (1968) found an inverse relationship between temperature and sensitivity. The positive correlation between temperature and sensitivity found in field experiments could result from the dominant influence of light intensity.

In fumigation experiments, Swain (1923) came to the conclusion that plants grown at 5° C or lower, were less sensitive to SO_2 than at higher temperatures. Sensitivity at temperatures above 5° C remained the same in the normal temperature range. Stoklasa (1923) mentioned no critical temperature limit, but states that sensitivity increases with an increase in temperature. Katz and Ledingham (1937) showed that plants became more resistant at temperatures above 38° C, whereas Setterstrom and Zimmermann (1939) found no differences in the sensitivity of buckwheat *(Fagopyrum esculentum)* to SO_2 at temperatures of 18° C and 40° C.

MacLean and Schneider (1971) determined that F accumulation and degree of injury were temperature dependent. After exposure to 4.7 μg HF/m^3 air for 104 h, gladiolus leaves showed 30% necrosis at the lowest temperature level of 16° C, as opposed to 70% and 62% at 20° C and 26° C respectively. The F accumulation at the highest temperature, however, was about 25% less than at the lower temperatures. In sunflowers *(Helianthus annuus)*, F accumulation in the leaves increased from 286 ppm, over 352 ppm to 473 ppm with increasing temperature. From this experiment it is obvious that, while pollutant uptake and degree of injury may be influenced by temperature, there are distinct species-dependent differences. Fluoride accumulation in above ground plant parts of alfalfa, orchard grass *(Dactylis glomerata)*, and endive *(Lactuca sativa* var. *romana)*, grown hydroponically in nutrient solutions with temperatures of 20° C, 25° C, and 30° C and exposed to 0.8 μg HF/m^3 air decreased with increasing temperature (Benedict et al., 1965).

After exposure of spruce *(P. abies)* to $^{35}SO_2$ at an outside temperature of −6° C, Materna and Kohout (1963) found counts of 16540 and 16880 impulses/min in the two youngest needle sets and showed that even outside the temperature range for active photosynthesis, sulfur uptake is possible. Alvik (1939) and Zeller (1951) have found that the lower temperature limit for apparent CO_2 assimilation for *Picea abies* is between −2° C and −3° C, while for *Picea mariana, Pinus silvestris*, and *Pinus nigra* var. *austriaca* the minimum temperature is −6° C (Freeland, 1944), and −5° C for *Pinus cembra* (Tranquillini, 1955). Katz (1949) found no SO_2 uptake by Douglas fir *(Pseudotsuga menziesii)* in winter.

Winter annuals still maintain apparent CO_2 assimilation even at temperatures below the freezing point. Sugar beet leaves, at −4.2° C, still absorbed 2.18 mg CO_2/h and 50 cm^2 (Lundegardh, 1927). Winter wheat *(Triticum sativum)*, winter barley *(H. vulgare)* and winter spinach *(S. oleracea)* in experiments carried out by Zeller (1951) near Stuttgart at the beginning of winter, still maintained active photosynthesis at temperatures from −2° C to −3° C. Respiration stopped at about −6° C to −7° C. That gas exchange still takes place at such low temperatures is a possible explanation for the fact that, in areas with SO_2 emissions, winter wheat and winter rye *(Secale cereale)* are more severely injured than summer grain varieties (Guderian and Stratmann, 1968).

2.5.1.3 Humidity

Turgor pressure in the leaves, a decisive factor for fully opened stomata and, therefore, for higher pollutant uptake rates, is largely dependent on moisture content of the air, as Lange et al. (1971), and others have shown on isolated epidermal tissues. In general, an increase in humidity increases the hazard to vegetation from air pollutants. Setterstrom and Zimmermann (1938) found, in experiments with SO_2 under otherwise constant conditions, that effects increased with increasing humidity. Differences in sensitivity, however, could first be demonstrated for humidity differences > 20%, a result which is in agreement with work done by Zimmermann and Crocker (1934). Zahn (1963b) has shown that resistance of currants became progressively greater as humidity dropped. For example, exposure to 0.8 ppm SO_2 for 8 h at 87% relative humidity caused severe leaf necrosis, but effects of 4.0 ppm SO_2 at 42% rel. humidity were only half as strong. At the lowest level of 29% rel. humidity, there was no visible injury.

From observations on various plant species, Zahn comes to the conclusion that the SO_2 sensitivity curve climbs slowly between 30 and 60% relative humidity and rapidly above 60%. Other investigations indicate that the critical point is above 70% (Swain, 1923; Haselhoff et al., 1932). Thomas and Hendricks (1956) report that sensitivity is reduced to one tenth when relative humidity drops from 100% to almost 0%. The extent to which SO_2 uptake and degree of injury are dependent on humidity is exemplified by the following growth chamber experiment with red clover (*T. pratense* var., *sativum*, Red Head). After exposure to 0.6 mg SO_2/m^3 air, for 80 h, plants grown under 90% relative humidity had accumulated a significantly larger amount of sulfur than those grown at 40% (Table 9).

This experiment primarily indicates a change in uptake rate. Therefore, different plant responses observed under various humidity conditions are essentially due to differences in pollutant accumulation.

The distinct dependance of S accumulation or leaf injury on relative humidity can be explained through stomatal action, as shown in numerous experiments with ozone. For example, a correlation has been found between stomatal pore size and leaf injury of the tobacco variety Bel W 3 and of bean (*P. vulgaris*) after exposure to ozone under different relative humidity conditions (Otto and Daines, 1969). No influence of humidity on ozone injury could be shown on ripe apples on which the lenticles are always open (Miller and Rich, 1968).

Of particular interest for growth chamber experiments are the results from Heggestad and Heck (1971) on the combined effects of relative humidity and light

Table 9. S accumulation and degree of injury of *Trifolium pratense* compared with humidity

Nonfumigated Control	Rel. humidity 40 %		Rel. humidity 90%	
S content in % dry matter[a]	Degree of injury[b]	S content in % dry matter	Degree of injury	S content in % dry matter
0.26 ± 0.01	8 ± 3	0.62 ± 0.04	50 ± 10	0.76 ± 0.03

[a] Average of six S analyses.
[b] Necrotic area in % of total leaf area; average of six plants.

intensity. Effects of 60 and 80% relative humidity on ozone injury did not differ under 30000 lux, but led respectively to 66 and 79% leaf necrosis under 20000 lux.

The importance of free water in the form of dew or rain droplets on leaves is still not entirely clear. Etching and acid burning have been observed on plants after rain in areas with high SO_2 concentrations (Guderian and Stratmann, 1962). Pollutants dissolve in the water film and rain drops. Through evaporation, the pollutant concentration is raised and necrosis can occur, especially on the tips of hanging leaves. Sensitive terminal leaves, particularly of rosaceous plants, also exhibit burn symptoms.

Observations over several decades have shown that in damp, rainy years, pollutant injury is more severe than in dry years with little precipitation (Wislicenus and Neger, 1914; Stoklasa, 1923). This phenomenon is due most likely to increased plant sensitivity, as in Western Europe in wet years the wind comes mainly from the west and, therefore, lower pollutant loading can be assumed. Because of higher relative humidity, photochemical pollutants have more severe effects on the east coast of the USA than on the arid west coast in California (Heggestad et al., 1964).

2.5.2 Edaphic Factors

2.5.2.1 Soil Moisture

In fumigation experiments, Zimmermann and Crocker (1934) have shown that wilted plants react less to SO_2 than plants under full turgor pressure. Katz (1937) and Katz and Ledingham (1937) came to the same conclusion after a series of experiments in soils with differing moisture content, but with otherwise comparable external conditions. As the wilting point was approached, sensitivity decreased greatly. Additional water above the point which guarantees optimal growth, however, has little effect on sensitivity, as shown in experiments with HCl. Van Haut (1961) and Zahn (1963 b) found reduced reactions to SO_2 on plants that had been deprived of water but had not yet begun to wilt. Setterstrom and Zimmermann (1939) studied the effects of soil moisture on the sensitivity of buckwheat *(F. esculentum)* before and during fumigation. Plants optimally supplied with water before fumigation exhibited much more extensive leaf necrosis than plants with a succulent habit that had been deprived of water before fumigation, even when moisture content of all soils was the same during fumigation. Experiments with various moisture contents confirm the findings of Katz (1937) and Katz and Ledingham (1937) mentioned above.

Results of experiments on the effects of soil moisture on sensitivity to SO_2 are comparable to those found under exposure to ozone. Oertli (1959) was able to show that resistance to ozone was raised when soil moisture dropped and osmotic pressure increased. His assumption that the weaker reactions were due to a reduction in the degree of stomatal opening was verified with porometer measurements (Seidman and Robert, 1963; Seidman et al., 1965). MacDowall (1965) found that plants deprived of water several weeks before fumigation only opened their stomata during the morning hours even with an optimal supply of water. Whether or not water supply changes the degree of plant resistance can only be proven with an exact analysis of the amount of pollutant taken up by the plant.

The results described above show that effects of various moisture contents in soils can be compared with effects of the moisture content of the air. In both cases, stomatal pore size changes and, therefore, the rate of uptake of air pollutants is altered (see Sect. 2.5.1.3).

2.5.2.2 Nutrition

The supply of nutrients, or substances indispensible for optimal growth and development, is also a determining factor for plant sensitivity to atmospheric pollutants (Stoklasa, 1923; Heck et al., 1965; Kisser, 1966). The number of experimental investigations, however, is still quite small and presents an incomplete and controversial picture.

Alfalfa *(M. sativa)* and buckwheat *(F. esculentum)* to SO_2, had a higher sensitivity when grown in nutritionally poor soils than in composted soils (Setterstrom and Zimmermann, 1939). Plants grown in a diluted (1/10) nutrient solution were more sensitive than those in the normal solution. As Thomas et al. (1943) have also shown, the sulfur content of the nutrient substrate had no effect on the SO_2 sensitivity of alfalfa. Leone and Brennan (1972), on the other hand, proved that leaf injury on tobacco and tomato grown in sand culture increased with an increase in sulfur. The degree of injury could be correlated with S accumulation, both from the nutrient solution and from the air. The reduced SO_2 uptake of the sulfur-deficient plants could result from a reduction in stomatal pore size, as was shown by Desai (1937) on K, P, and N-deficient plants. The degree of injury to strawberries grown in quartz sand and supplied with different nutrient solutions decreased as the amount of nitrogen was increased (Kisser, 1966).

Van Haut and Stratmann (1960, 1970) found similar responses. On ornamentals, such as *Gloxinia* and *Pelargonium*, effects on quality and number of blossoms could be reduced through fertilizers high in P and K. Zahn (1963b) showed that resistance of dicots was increased by raising the amounts of mineral nitrogen. With winter barley, however, there was no connection between sensitivity and nitrogen supply, and fertilized oat plants were more sensitive than nonfertilized plants. Guderian (1971) demonstrated a distinct increase in resistance through addition of nitrogen in experiments with winter wheat, summer rape *(B. napus* var. *oleifera)*, and sunflowers *(H. annuus)* in various soils. For example, fertilized wheat showed only slightly more leaf injury than nonfertilized controls, in spite of sulfur accumulation from exposure to SO_2 that was three times higher than the controls. On the other hand, reduced injury to spruce, fertilized with ammonium sulfate, was due to lower SO_2 uptake.

The same results have been reported by Materna and Kohout (1967) for spruce *(P. abies)* fertilized with urea. Figure 22 presents the effects of various amounts of Ca, P, and K on plant reactions to SO_2. Experiments were carried out with a sandy loam soil with a pH (KCl) of 4.2 and lactate soluble P_2O_5 and K_2O amounts of 2 and 4 mg/100 g soil, respectively (Guderian, 1971).

There is a distinct connection between the amount of lime and the degree of injury to red clover and winter barley. The highest amounts of $CaCO_3$—200 and 400 mg/100 g soil—still reduced injury to barley but not to red clover, a fact which is probably dependent on the different Ca requirements of the two plants

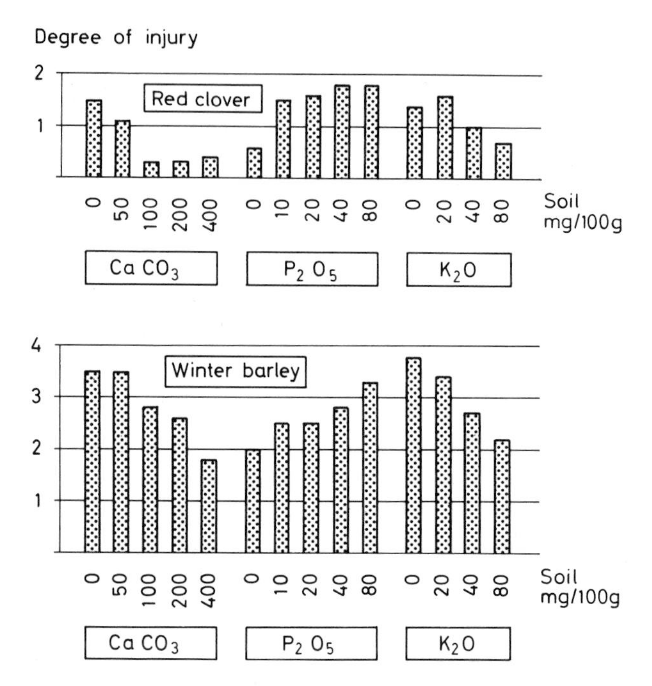

Fig. 22. Influence of the step-wise addition of mineral fertilizer on the extent of injury on red clover and winter barley (*O* = Control)

(Klapp, 1967). Through comparisons of sulfur accumulation and degree of injury it can be seen that balanced Ca levels in the soil increase resistance; SO_2 uptake increased only slightly with an increase in Ca. Potassium had no effect on sulfur accumulation, but, like, lime, significantly increased resistance of leaf tissue. With an increase in phosphorous, sulfur accumulation and extent of leaf injury increased on red clover, as well as on winter barley.

Enderlein and Kästner (1967), in experiments on SO_2 sensitivity of pine *(P. silvestris)* in a soil deficient in N, P, K, Ca, and Mg, found that the slightest injury occurred when all nutrients were available. The nutrient most effective in reducing injury proved to be nitrogen, as field experiments in the Dübener Heath have also shown. Plants supplied with P, K, Mg, and Ca exhibited slightly less needle injury than nonfertilized controls. Materna (1962, 1963) found a generally higher productivity in experiments with basic pulverized rock, lime, phosphorous, and nitrogen, but an increase in resistance under practical conditions could not be proven. In a chronically polluted area Trillmich (1969) showed that fertilizing with PKCaMg fertilizer and a second treatment with a NPKCaMg fertilizer positively influenced growth and development of a mixed stand of broad-leaved and coniferous trees.

Similar to the effect on sulfur uptake, addition of nitrogen reduced accumulation of fluoride in spruce *(P. abies)* needles (Table 10). The extent of needle injury was significantly less than the nonfertilized controls only on plants supplied with 9 g $(NH_4)_2SO_4$/8 l pot. An increase in plant sensitivity after addition of large

Table 10. Degree of injury and F accumulation in *Picea abies* exposed to HF compared with supply of nitrogen

Treatment	F content in ppm dry matter[a]	Degree of injury[b]
Nonfumigated control		
Without $(NH_4)_2SO_4$	15, 16 } $\bar{x}=16$[b]	—
9g $(NH_4)_2SO_4$/pot	16, 16 } $\bar{x}=16$	—
18g $(NH_4)_2SO_4$/pot	15, 17 } $\bar{x}=16$	—
270 h · 5.4 μg HF/m³ air		
Without $(NH_4)_2SO_4$	47, 49 } $\bar{x}=48$	3.65
9g $(NH_4)_2SO_4$/pot	41, 43 } $\bar{x}=42$	2.40
18 g $(NH_4)_2SO_4$/pot	26, 27 } $\bar{x}=27$	3.15

[a] Average of two mixed samples of five plants each.
[b] Average of two evaluations of 10 plants each of spruce clone GA 22.

Table 11. Degree of injury and F accumulation in *Hordeum vulgare* compared with supply of P, K, and Ca

Treatment	F content in ppm dry matter	Degree of injury[a]
Nonfumigated control	80 ± 10[b]	
192 h · 12.3 μg HF/m³		2.6 ± 0.4
without P_2O_5	1960	
10 mg P_2O_5/100 g soil	2230	2.8 ± 0.5
20 mg P_2O_5/100 g soil	2420	2.5 ± 0.3
40 mg P_2O_5/100 g soil	2430	2.4 ± 0.3
80 mg P_2O_5/100 g soil	3390	2.2 ± 0.4
Nonfumigated control	90 ± 30[b]	
192 h · 12.3 μg HF/m³		1.5 ± 0.3
without K_2O	2410	
20 mg K_2O/100 g soil	2780	1.8 ± 0.4
40 mg K_2O/100 g soil	3760	1.6 ± 0.2
80 mg K_2O/100 g soil	3590	1.6 ± 0.5
Nonfumigated control		
192 h · 12.3 μg HF/m³		2.6
50 mg $CaCO_3$/100 g soil		2.1
100 mg $CaCO_3$/100 g soil		2.0
200 mg $CaCO_3$/100 g soil		1.4
400 mg $CaCO_3$/100 g soil		0.9

[a] Average of two evaluations of five pots each.
[b] Average of five samples.

amounts of nitrogen has also been reported for SO_2. Experiments were carried out with a sandy loam soil with a pH (KCl) of 5.3 and lactate soluble P_2O_5 and K_2O amounts of 8 and 14 mg/100 g soil, respectively. Addition of fertilizer occurred once a week for three weeks, the last application was one week before exposure to HF.

After exposure to HF, spinach in soils treated with nitrogen, added either as ammonium sulfate or as calcium nitrate, also had a lower F accumulation than in other treatments. The fertilized plants also showed significantly less injury than nonfertilized plants.

The connection between F accumulation and leaf injury on the winter barley variety Peragis 12 Melior after the step-wise addition of P, K, and Ca is shown in Table 11.

In these experiments, carried out on a loamy sand soil with a pH (KCl) of 4.2 and lactate soluble P_2O_5 and K_2O amounts of 1 and 9 mg/100 g soil, respectively, the fluoride accumulation increased with an increase in P and K. The resistance-increasing effects of P and K could be clearly seen in the degree of injury, which remained the same or decreased slightly. The extent of injury was greatly reduced through the addition of $CaCO_3$. Comparable results were found by Pack (1966) on tomato in a sand culture experiment. Plants deprived of Ca had not only extensive leaf injury, but also smaller and sometimes seedless fruits. It is assumed that fluoride acts on the calcium metabolism during fructification. MacLean et al. (1969) describe effects on tomato deprived of Ca, Mg, and K that agree with results mentioned above.

Brennan et al. (1950) found that a deficiency, as well as over-supply, of N, P, and K caused a reduction in fluoride uptake of tomato plants, whereas an optimal supply of these nutrients promoted fluoride injury. These findings may be due to the fact that Brennan et al. (1950) chose HF concentrations that were unrealistic and caused severe acute injury. During the uptake of large amounts of a pollutant in a short time, gradual resistance differences have no effect. Under such conditions, the plants with the highest pollutant accumulation are most severely injured. In contrast to this are the results of Applegate and Adams (1960), which show that fluoride uptake was increased in bean seedlings deficient in P, K, or Fe. MacLean et al. (1969) carried out HF fumigations on tomato plants grown in sand culture either in a complete nutrient solution or in solutions deficient in Ca,Mg, and K. Fluoride reduced growth and increased the deficiency symptoms on basal leaves of plants grown under Mg deficiency. Fluoride accumulation was reduced through Mg deficiency. Effects of HF on Ca-deficient plants were characterized by reduced growth, a general increase of chlorosis, and increased necrosis of young leaves. Fluoride accumulation was not influenced. Potassium deficiency increased fluoride accumulation in leaves and stems and increased the extent of necrosis on apical leaves.

2.5.3 Influence of Stage of Development and Leaf Age

For the determination and evaluation of developmental differences in plant sensitivity, it is important to differentiate between the process of development of the individual plant and of single plant organs. Both forms of development determine type and extent of plant responses to a given pollutant.

2.5.3.1 *Stage of Development*

During the course of their development, higher plants respond quite differently to air pollutants. Short-lived plants show a strong degree of resistance during early stages of development. Grain plants are most resistant during the early stages from sprouting to the two-leaf stage (van Haut, 1961). They are most sensitive in the three-leaf stage and afterward they become more resistant. During shooting and especially just before the appearance of the inflorescense, leaves become more sensitive and remain so until ripening, as has been shown in fumigation experiments with HF and H_2SiF_6 on sorghum (Hitchcock et al., 1963). Winter barley *(H. vulgare)* was more sensitive to HF during the two-leaf stage and tillering than during the development phases between these two stages (Guderian, 1969).

Pollutant action in the "critical stages of development" from grain formation to ripening can cause severe negative effects on yield. This is understandable, as during this physiologic phase of "switching over" to generative production most of the grain building substances are transported into ears (Heyland, 1959). Leaf injury during tillering under otherwise good growth conditions, however, has little effect on grain yield. In dicots, for example bush bean *(P. vulgaris nanus)*, the critical stage of development is also in the time between blooming and early ripening. Injury to the cotyledons can also have a severe effect on yield. Observations in the field have shown this to be particularly true for beets exposed to gaseous fluoride compounds.

Different changes in sensitivity, due to various stages of development of single plant species, can result in a shifting or even reversal of resistance relationships of various plants to one another, as Guderian (1966b) has shown in fumigation experiments with oat cultures overgrown with wild mustard *(Sinapis arvensis)*; (Fig. 23).

Investigations on pollutant uptake during various stages of development have shown that a certain parallel exists between pollutant uptake and metabolic production (Guderian et al., 1969; Guderian, 1970). Sulfur accumulation, as well as fluoride accumulation, increased with an increase in growth. This also appears to be the case with chloride accumulation.

Oats in the four leaf stage, fumigated with 0.45 mg HCl/m^3 air for 32 h, showed an increase in chloride from 1.68 to 2.60% in dry matter. Plants in the

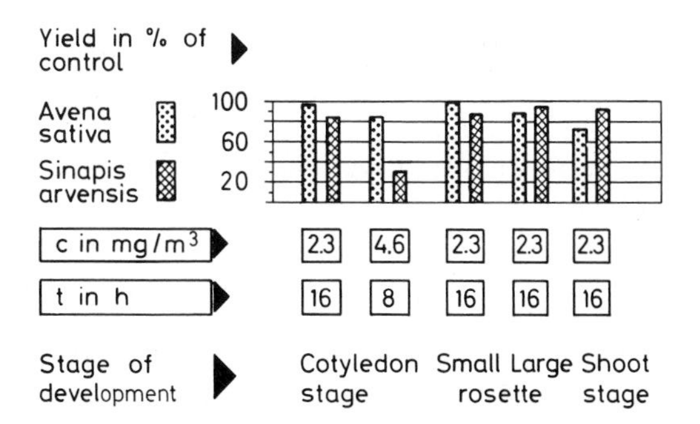

Fig. 23. Effect of the time of exposure to SO_2 on growth of cultivated and wild plants based on an example of an oat culture interspersed with wild mustard

tiller stage, however, showed only an increase from 2.00 to 2.37%. Dry-matter production of the younger plants tripled during the time of the experiment, whereas the older plants showed an increase of only 1.5 times.

Injury, in the form of leaf necrosis, or reduction of growth, did not correlate with sulfur accumulation. From the vegetative development phases studied, the younger stages proved to be the most resistant, that is, in spite of a usually higher sulfur accumulation there was less foliar injury.

There is comparatively little information available on the sensitivity of perennials during various phases of development. Based on field observations, Wentzel (1963) has shown that a certain sensitivity of conifers begins in the late sapling stage at the time of cumulative growth and remains until early maturity. Injury during this time can lead to severe reductions in growth and an opening of entire forest stands (Wentzel, 1962; Knabe, 1970a; Materna et al., 1969). Under chronic exposure to SO_2, seedlings of spruce *(P. abies)* and pine *(P. silvestris)* showed a degree of resistance similar to oak and beech seedlings. In later stages of development, however, oak and beech became more resistant (Guderian and Stratmann, 1968).

Whether this is completely or only partly due to inherent resistance is unknown. Tall saplings and trees in the forest canopy are exposed to a higher air exchange. This results in different pollutant accumulation rates at the same pollutant concentration, as reported by Knabe (1968) for various forest strata in polluted areas. Fluoride uptake increased with an increase in plant height. In species with a long rotation time, this leads to summation effects which cause a growth habit typical for the polluted area.

Pollutant injury to highly sensitive conifers causes a decrease in needle growth, a flattening of the canopy, a gradual opening of the stand through dryness, and, finally, dissolution of the entire stand. Throughout this process there are distinct differences between death rates due to chronic and acute injury (Pelz and Materna, 1964). Pollutant effects on seed and stone fruit trees include a decrease in leaf size, early leaf fall which, in connection with metabolic disturbances, can lead to a decrease in twig and bud growth in the following year, and influences on maturity of fruits and wood (Auersch, 1967; Guderian, 1969). Reduced lateral and branch growth hinders the development of a robust canopy, which ensures high yields over a long period of time. Early senescence may also occur and leads to a shortening of the yield period. Because of severe growth retardation on the exposed side, asymetrical canopy growth occurs over a period of years.

Differences in resistance of perennials are particularly great during rhythmic phases of activity. There is little risk from air pollutants to deciduous trees and shrubs during the winter. Evergreen plants also show less injury during dormant phases than during active growth, largely as a result of reduced gas exchange.

2.5.3.2 Leaf Age

Of all plant organs the leaf is the most sensitive to the air pollutants studied here, as well as to a large number of other external factors. This sensitivity rests on the fact that the major portion of important physiologic processes are concentrated in the leaf, which acts as the center of variability or plasticity of the organism. Therefore, the leaf, with its various stages of development, is an especially good indicator for particular atmospheric pollutants.

The sensitivity of the leaf, determined through leaf age, varies with type and concentration of a given pollutant. Observations are reported in the first books on air pollutant effects (von Schröder and Reuss, 1883; Haselhoff and Lindau, 1903) in which leaf age is correlated with SO_2 sensitivity. Fumigation experiments in small greenhouses (Zimmermann and Crocker, 1934; Setterstrom and Zimmermann, 1939) and in growth chambers (van Haut, 1961; van Haut and Stratmann, 1970) have shown that younger leaves are generally more resistant than more fully expanded leaves. Leaves in stages of full growth are more severely injured by acute concentrations than younger or older leaves, as can be seen in Figure 24 (see p. 124). As indicated by these experiments, and many others with short exposures to high SO_2 concentrations, the extent of leaf injury is more dependent on physiologic age of the leaf than on the actual age.

The relationships in the degree of injury to leaves of different ages vary greatly with pollutant concentration (Guderian, 1970). With long-term exposures to low concentrations, the older leaves are usually injured before the younger leaves, as experiments with apple, pear, beet *(Beta vulgaris)*, and broad bean *(Vicia faba)* have shown.

Figure 25 presents results of an experiment in which common alder *(Alnus glutinosa)* was exposed to SO_2 concentrations of 3 mg, 5 mg, and 6 mg/m^3 air. After three days under 3 mg SO_2/m^3 air, the oldest leaves, with the weakest photosynthesis and lowest SO_2 absorption, were injured before the younger leaves. With an increase in concentration, effects extended to the younger leaves until, finally, under the highest concentration, the extent of injury of freshly expanded leaves could be correlated with sulfur accumulation and photosynthesis.

In these experiments, in agreement with many others, the highest sulfur accumulation was found in the almost or just fully expanded leaves, which also had the highest rates of photosynthesis. From comparison of the extent of injury with sulfur accumulation, it is clear that younger leaves, contrary to earlier reports (Zimmermann and Crocker, 1934), have a specifically higher resistance. This resistance is manifested in reduced leaf injury under long-term exposures to low concentrations, as opposed to short-term exposures to acute concentrations. Under high concentrations, these differences in resistance could not be seen because of the higher SO_2 uptake of the newly opened leaves.

Pollutant uptake and degree of foliar injury are dependent not only on leaf age, but also on age of the entire plant (Guderian, 1970). SO_2 uptake of sunflower *(H. annuus)* leaves of the same age decreased with an increase in plant age. This occurs because young leaves of older plants have lower gas-exchange rates than young leaves of younger plants. Comparisons of sulfur accumulation and extent of injury also show that leaves which are physiologically more active are also more resistant.

Plants such as tobacco or broad bean *(V. faba)*, which grow in length, also exhibit the described age-dependent leaf sensitivity. Throughout development, the lower leaves become successively more resistant than younger leaves higher up

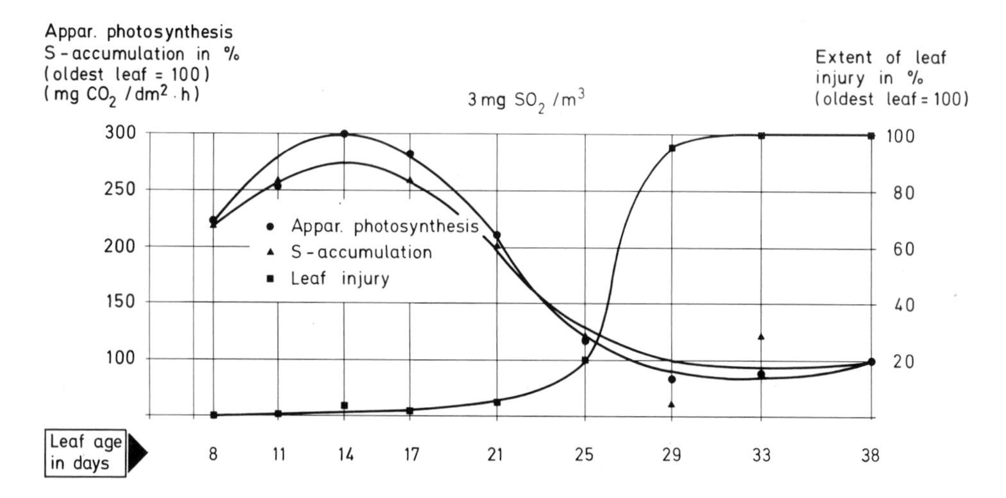

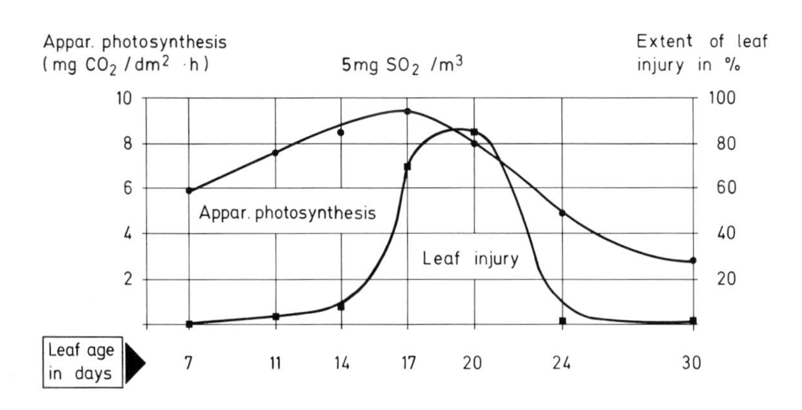

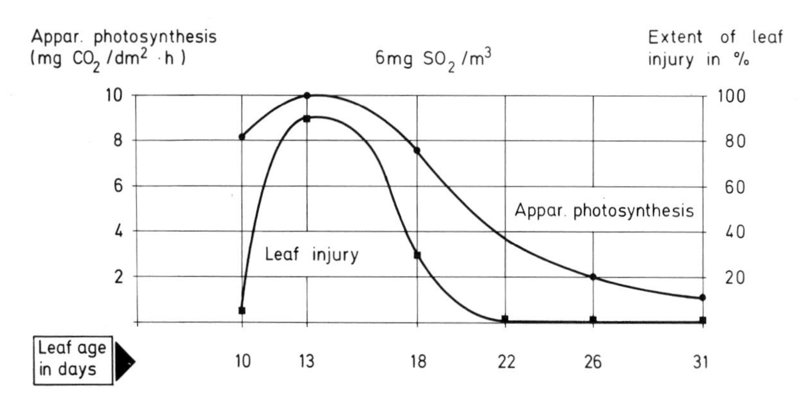

Fig. 25. Extent of leaf injury and sulfur accumulation in alder after exposure to various SO_2 concentrations compared with leaf age and apparent photosynthesis

the stem (van Haut and Stratmann, 1970). Plants with a rosette growth habit, such as beet, celery *(Apium graveolens)*, and lettuce *(L. sativa)*, show injury first on the outside leaves rather than on younger innermost leaves. With age, sensitivity moves progressively toward the center of the plant.

After exposure to HF, it is common that symptoms of injury (Fig. 26, see p. 125) are first seen on very young leaves that are still expanding, as experiments on sugarbeet, poplar (*Populus* sp.), common alder *(Alnus glutinosa)*, Norway maple *(A. platanoides)*, birch *(Betula pendula)*, dahlia *(Dahlia pinnata)*, and carnation *(Dianthus caryophyllus)* have shown (Guderian et al., 1969). Developing leaves of sugarbeet showed severely reduced growth without visible signs of injury. Necrosis occurred on young developing needles of Ponderosa pine, although fluoride accumulation was lower than in needles from previous years (Adams et al., 1956).

Such effects on young developing leaves do not appear to be a result of high fluoride accumulation, since fluoride accumulation increased with leaf age in other species, such as Marrowstem kale *(B. oleracea* var. *medullosa)*, kale *(B. oleracea* var. *acephala)*, and spinach. Young leaves of certain dicots show a particular sensitivity which can be related to the effects of fluoride on the formation of magnesium containing substances and, therefore, on biosynthesis of leaf pigments (McNulty and Newman, 1961). Under SO_2, a breakdown of chlorophyll to pheophytin usually occurs (Dörries, 1932; Rao and LeBlanc, 1966; Arndt, 1971).

Contrary to effects of HF, but similar to those of SO_2, newly expanded leaves show the most extensive injury from exposure to HCl (Fig. 27, see p. 126).

There is no distinct connection between chloride accumulation and leaf age (Table 12). In nonfumigated control plants, a definite increase in chloride content with advancing leaf age can be seen, but deviations between the various leaf stages are greater than with fluoride and considerably greater than with sulfur.

The described trends can also be observed in conifers: a definite decrease in sulfur accumulation with increasing needle age (Materna, 1965a; Guderian, 1970), almost constant fluoride accumulation in all needle sets, regardless of age, and similar effects with HCl, as shown in experiments with *Picea abies* and *Pinus silvestris* (Table 13).

It is still an open question as to why uptake of HF and HCl, as opposed to SO_2, does not correlate with gas-exchange activity of leaves. It is assumed that greater amounts of Cl and F, after solution in the water film of the leaf blade, enter the mesophyll of younger leaves through the cuticle, which is permeable for dissolved substances (Franke, 1967). Brewer et al. (1960b) by spraying concentrated solutions of HF, NaF, NH_4F, H_2SiF_6, and CaF_2 on leaves, demonstrated that fluoride can enter the leaf mesophyll in this way. HF, NaF, and H_2SiF_6 were most quickly absorbed. The differences in uptake rate are assumed to depend on the size of the cations. As experiments on hypostomatic leaves of *Fatshedera lizei* Guillaum have shown, fluoride from HF concentrations in the air, can be taken up directly through the epidermis (Grafe, 1973). Fluoride content of leaves treated on the underside with polyvinyl alcohol, which is nonpermeable to HF, and exposed to 31 μg HF/m^3 air for 24 h, increased by 11 ppm in dry matter, whereas leaves treated on the top side had an increase of 44 ppm in dry matter. In a fumigation experiment with HCl, no significant differences in chloride accumula-

Table 12. Cl accumulation in currant *(Ribes rubrum)*, grape *(Vitis vinifera)*, apple *(Malus communis)*, and cherry *(Prunus avium)* leaves of different ages exposed to HCl (in % dry matter)

Leaf number[a]	Cl content Control	After HCl exposure	Cl accumulation
Currant		48 h · 0.45 mg HCl in the time from 2–6 July	
1– 4	0.44	0.81	0.37
5– 8	0.40	0.67	0.27
9–12	0.36	0.61	0.25
13–16	0.25	0.63	0.38
17–19	0.19	0.57	0.38
> 19	0.10	0.63	0.53
Grape		48 h · 0.45 mg HCl in the time from 2–6 July	
1– 4	0.30	0.65	0.35
5– 8	0.17	0.39	0.22
9–12	0.10	0.36	0.26
> 12	0.09	0.64	0.55
Apple (Golden Delicious)		32 h · 0.42 mg HCl in the time from 10–13 July	
1– 4	0.15	0.40	0.25
5– 8	0.06	0.36	0.30
9–12	0.08	0.37	0.29
13–16	0.04	0.31	0.27
> 16	0.04	0.28	0.24
Cherry		32 h · 0.42 mg HCl in the time from 10–13 July	
1– 4	0.16	0.38	0.22
5– 8	0.14	0.32	0.18
9–12	0.05	0.26	0.21
13–16	0.04	0.21	0.17

[a] The highest numbers indicate the youngest leaves.

Table 13. Cl accumulation and degree of injury in spruce *(Picea abies)* and pine needles *(Pinus silvestris)* of different ages exposed to HCl (in % dry matter)

Needle age	Control	70h · 0.22mg HCl (6–14 July)	120h · 0.25mg HCl (6–20 July)
Spruce			
One-year-old	0.06	0.19	0.28
Two-year-old	0.07	0.18	0.28
Pine			
One-year-old	0.10	0.24	0.31
Two-year-old	0.15	0.28	0.36

tion were observed between apple (Golden Delicious) leaves that were sprayed every hour and those that were kept dry.

Translocation can be excluded, at least for fluoride, as a factor which leads to changes in pollutant relationships in leaves of different ages. For example, no transport of fluoride from older to younger leaves, which at the time of fumigation

Table 14. Translocation of fluoride from two-year-old needles of *Picea abies* into new growth

Treatment	Needles from 1969 F content in ppm dry matter[a]		Needles from 1970 F content in ppm dry matter
	29 Sept. 1969	22 April 1970	13 June 1970
Control	16.6±0.6	25.6±2.1	7.7±0.2
72h · 26.8 µg HF/m³ air	66.6±6.7	50.6±2.1	8.2±0.2[−b]
108h · 26.8 µg HF/m³ air	77.3±2.0	68.6±4.7	8.7±0.5[+b]

[a] Average of five samples from each of two plants.
[b] − = not significant, + = significant at the 95% level compared with nonfumigated control plants.

were not yet fully differentiated, could be found in Marrowstem kale *(B. oleracea* var. *medullosa)* or in broad bean *(V. faba)*. This agrees with observations made by Hitchcock et al. (1963) on sorghum varieties. Translocation of fluoride from apical to basal sections of a spruce crown could not be proven. Benedict et al. (1964), with the help of chemical analyses, found no translocation of fluoride from leaves into stems or roots. A slight translocation effect was found only on new growth of *P. abies* clone GA22, which was fumigated during the preceeding year (Table 14).

2.5.4 Specific, Varietal, and Individual Resistance

Single plant species, varieties, cultivars and individuals of a species react differently to a given air pollutant. There is no absolute resistance to gaseous air pollutants as some plants have against certain phytopathogens (Gäumann, 1951; Fuchs and Rosenstiel, 1958; Grossmann, 1970). Plants have neither inherent defense mechanisms (Anexie) nor induced defense mechanisms (Apergie) which hinder uptake of pollutants or effects from incorporated pollutants. Injury to all plant species occurs through action of a definite effects product $(c \cdot t)$, composed of concentration (c) and exposure time (t). The degree of resistance, expressed as species and individual-specific reactions, depends on genetic factors, stage of development, and environmental factors, and varies within a broad range.

According to Levitt (1972), resistance of an organism to deleterious external factors is composed of two components, "stress avoidance" and "stress tolerance." Stress avoidance includes mechanisms that influence pollutant uptake or toxicity. Under stress tolerance are mechanisms that influence the capacity of an organism to withstand effects of air pollutants.

Pollutant uptake can be reduced through specific morphologic, anatomic, physiologic, and ecologic characteristics (Türk et al., 1974). A reduction in toxicity is dependent on type and rate of biochemical and physiologic reactions, which in turn vary with the environmental factors soil, climate, and nutrition. Inherent resistance varies according to rhythmic activity that is dependent on time of year and stage of development of the plant.

Rohmeder et al. (1962, 1965), working with spruce *(P. abies)* and pine scions *(P. silvestris)* of phenotypically resistant stocks from polluted areas, found that resistance to SO_2 and HF is genetically controlled. This was confirmed by Schönbach et al. (1964) in their work on larch crosses, and Dochinger et al. (1965a, b) came to similar conclusions in investigations of "chlorotic dwarfing" from ozone injury to white pine *(P. strobus)*. The inherent differences in resistance are believed to result from biochemical differences in the protoplasm. This assumption is backed up by Brennan and Halinsky (1970), who found that certain species and varieties show differences not only in the degree of sensitivity but also in the type of external injury symptoms.

Inverse relationships between degree of injury of certain species and varieties and pollutant accumulation in leaves also provide proof of species-specific reactions. For example, highly sensitive tulip and gladiolus varieties show injury with a fluoride accumulation of only 10–20 ppm in dry matter, whereas cabbage often remains uninjured with an accumulation of 2000 ppm and cotton shows no injury even at 4000 ppm in dry matter (Thomas and Alther, 1966). Guderian et al. (1969) found little correlation between fluoride uptake and degree of leaf injury on various grass and clover varieties. There was also no definite connection between fluoride content and tip necrosis of several tulip varieties, which confirms the findings of Hitchcock et al. (1963) on sorghum varieties. Gladiolus, on the other hand, usually shows an inverse correlation between fluoride content and leaf injury (Compton and Remmert, 1960; Johnson et al., 1950; Hitchcock et al., 1962). The highest fluoride levels were found in the least injured areas.

The reasons for these different resistance responses are, for the most part, still unknown. Pelz and Materna (1964) could find no significant differences in glycoside or amino acid content between phenotypically resistant and sensitive spruce trees. There seems to be a relationship, however, between seasonal amounts of volatile substances in the needles, cell sap concentrations, and frost resistance in resistant individuals. Needles of resistant trees also had higher levels of potassium in spring than the injured reference trees.

Zimmermann and Hitchcock (1956) found no connection between sensitivity to HF of 25 different species and content of SiO_2, CaO, and Fe. In personal investigations, no connection was found between resistance and pH of the cell sap or aluminum content of the leaves. Grape leaves supplied with water that contained 5 or 10 ppm $Al(NO_3)_3 + C_4H_6O_6$ showed significantly less injury than those supplied only with distilled water during fumigation. Reusmann et al. (1971) have shown that fluoride ions can be bound with the addition of Al ions and precipitate as apatite after addition of Ca and PO_4 ions.

Specific, varietal, and individual differences in resistance, however, come only in part from biochemical differences. Determining factors are the levels of accumulated pollutants, especially in the leaves and cell organelles, and translocation of pollutant components in the plant. Rohmeder et al. (1962, 1965), as well as Pelz and Materna (1964) found lower amounts of sulfur in needles of phenotypically resistant spruce trees than in needles of sensitive trees. Physiologically more active trees with high photosynthesis rates and greater lateral growth proved to be more sensitive. Engle and Gabelman (1966) determined that resistance of onion *(Alium cepa)* was dependent on a particular dominant gene pair. In the presence of

Table 15. Changes in F content of various leaf sections of the gladiolus variety Snow Princess after exposure to HF (Hill, 1969)

Time after start of fumigation	F content in ppm dry matter		
	Leaf tip	Mid section	Basal section
1 h · 890 μg HF/m^3 air			
2 h	71	87	92
5 h	94	115	68
45 h	159	104	45
24 h · 40 μg HF/m^3 air			
1 day	132	97	34
2 days	188	83	26
10 days	224	71	27

ozone, guard cells of resistant plants lost turgor pressure, as a result of changes in permeability, causing closure of the stomata, which hindered further pollutant uptake. Stomata of sensitive plants remained open.

The importance of secondary translocation for the negative effects of particular pollutants can be seen in the following examples. Gladiolus, which reacts to HF similarly to other members of the Iridaceae and Liliaceae, had fluoride levels in the leaf tips which were 10 times higher than in basal portions of the leaf (Compton and Remmert, 1960). After long-term exposures to low concentrations even differences of 1:100 were found (Hill, 1969).

From Table 15 it is evident that these differences come less from variations in absorption rates of the different plant parts than from secondary translocation of the pollutant toward the leaf tip.

Comparatively small differences in the fluoride content of various leaf parts were found in fumigation experiments carried out by the author with a clone of balsam poplar *(P. tacamahaca)* which primarily shows injury on leaf margins (Table 16).

Within seven days of fumigation, there was a significant reduction in fluoride levels, but the relationship between marginal and medial portions of the leaf remained the same. On Marrowstem kale *(B. oleracea* var. *medullosa)*, however, no significant differences could be found in fluoride content of various leaf parts. This was true for natural fluoride levels, as well as fluoride levels resulting from exposure to HF. That fluoride is not translocated to the margins of kale leaves contributes to the fact that these very resistant plants will only be injured when grown in very heavily polluted areas.

The high, inherent degree of resistance of kale becomes clear when compared with the reactions of balsam poplar. Poplar leaves had 2 mm-wide necrotic lesions on the leaf margins, whereas kale, with fluoride levels which were four times higher, remained free of injury. Fluoride levels in kale leaves also indicate that pollutant uptake rates are essentially the same for all sections of the leaf. The assumption of Romell (1941), that HF uptake rates are higher at the leaf margins because of disturbances in diffusion relations in the leaf, apparently does not hold true.

Table 16. F content in various leaf sections of Balsam poplar *(Populus tacamahaca)* and Marrowstem kale *(Brassica oleracea)* before and after exposure to HF

Treatment	F content in ppm dry matter	
	Middle section without midrib	Leaf margin
Balsam poplar		
Control	13 ± 0[a]	31 ± 4[b]
160 h · 13.1 μg HF/m^3 air	220	308
7 days after exposure	172	246
Marrowstem kale		
Control	14 ± 5	14 ± 3
286 h · 22.1 μg HF/m^3 air	1158 ± 5	1165 ± 6
14 days after exposure	1013 ± 30	1038 ± 29

[a] Average of double analyses consisting of three samples.
[b] Average of double analyses consisting of five samples.

Similar to injury from HF, necrosis of leaf margins predominates after exposure to HCl. In contrast to previous assumptions, however, this necrosis apparently is not a result of higher chloride levels in the leaf margins, but of the higher sensitivity of these tissues (Table 17).

Natural chloride content of lilac leaves in a 3 mm-wide marginal area, middle area, and medial section without midrib, with values of 0.20%, 0.19%, and 0.20% of dry matter respectively, were practically constant throughout the leaf. After exposure to HCl, no changes in these relationships occurred. Lilac and grape are species in which marginal necrosis occurs almost exlusively. There were no differences in the natural chloride content of grape leaves between a 5 mm-wide marginal area and the rest of the leaf blade. After exposure of grape to 0.68 mg HCl/m^3 air for 80 h, the leaf margins showed chloride levels which were distinctly

Table 17. Cl content in various leaf sections of lilac *(Syringa vulgaris)* and grape *(Vitis vinifera)* before and after exposure to HCl

Treatment	Cl content in % dry matter[a]		
	Leaf margin	Outer zone	Midsection
Lilac			
Control (before exposure)	0.20 ± 0.01	0.19 ± 0.01	0.20 ± 0.02
Control (after exposure)	0.22 ± 0.01	0.23 ± 0.02	0.23 ± 0.02
100 h · 0.18 mg HCl/m^3 air	0.41 ± 0.02	0.42 ± 0.03	0.40 ± 0.03
7 days after exposure	0.40 ± 0.03	0.43 ± 0.03	0.44 ± 0.04
Grape			
Control (before exposure)	0.04 ± 0.01		0.05 ± 0.01
Control (after exposure)	0.04 ± 0.01		0.04 ± 0.01
80 h · 0.68 mg HCl/m^3 air	0.69 ± 0.02		0.90 ± 0.09
7 days after exposure	0.54 ± 0.03		0.63 ± 0.04
14 days after exposure	0.54 ± 0.03		0.58 ± 0.03

[a] Average of double analyses of four samples.

lower than the rest of the leaf blade. Necrosis, however, occurred only on leaf margins. Within two weeks after exposure all leaf sections had attained the same chloride content. In general, it appears that plants that show differences in the natural content of a particular element between leaf margins and the rest of the leaf blade also show differences in these leaf sections after exposure to a pollutant.

Secondary translocation of sulfur in the leaf appears to be relatively unimportant for the action of SO_2. Accumulation in marginal areas can occur, however, as shown in analyses of leaf parts of pears and beets (Guderian, 1970). Marginal necrosis, especially on some of the Papilionaceae, observed after long-term exposures to low concentrations of SO_2, could be explained through such transport mechanisms (van Haut and Stratmann, 1970; Guderian and van Haut, 1970). Kisser et al. (1962) and Halbwachs (1963) have shown this transport to be a result of streaming toward the leaf margins and tips that is caused by a "sucking" action in the leaf margins. The causes of this sucking action remain to be clarified. From these few examples it can be seen that secondary transport, especially of fluoride, is a determining factor for type and extent of injury of particular plant species.

The degree of resistance of plants changes greatly with pollutant type. For example, tuberous and bulbous species and stone fruits, such as apricot and prune as well as geranium and corn, are exceptionally sensitive to HF, but relatively resistant to SO_2. Legumes, sensitive to SO_2, however, are quite resistant to HF (Zimmermann and Hitchcock, 1956; Guderian and Stratmann, 1968; Jacobson and Hill, 1970). For a more extensive discussion of plant resistance to SO_2 and HF, the reader is referred to summaries by Thomas (1961), Garber (1967), Guderian and Stratmann (1968), Wentzel (1968), Guderian et al. (1969), and van Haut and Stratmann (1970).

Resistance of plants to HCl has only recently been investigated. Therefore, the following resistance series, based essentially on fumigation experiments with HCl concentrations which cause acute injury, is presented (van Haut and Guderian, 1976). The degree of resistance of important agronomic plants has been divided into groups of very sensitive, sensitive, and less sensitive species (Tables 18 and 19). These groups provide the basis for determining effects of HCl with differential diagnostic methods. Suitability for cultivation should be based on effects on intended use, as is discussed in the following section.

Aside from inherent factors, degree of resistance varies with changes in development of the plant (see Sect. 2.5.3.1), with climate, and with soil conditions (see Sects. 2.5.1 and 2.5.2). Without doubt, some of the variations in the "resistance series" come from differences in growing conditions in the various study areas (Stoklasa, 1923; Haselhoff et al., 1932; Bredemann, 1956; Thomas, 1961; Garber, 1967; Bolay and Bovay, 1965).

The cultivation of plants in polluted areas, in spite of their unsuitability, is usually due to a lack of consideration of external factors that increase plant sensitivity. For example, red beech (*Fagus silvatica*) is much more resistant on limey soils than on nutrient-deficient sandy soils (Wentzel, 1968). Elm (*Ulmus carpinifolia*), however, proved to be one of the most resistant deciduous species when grown on soils high in lime and one of the most sensitive on unsuitable soils.

The importance of species-specific differences in resistance on composition of plant communities is most apparent under conditions of a pollutant concentra-

Table 18. HCl resistance of deciduous and coniferous trees based on leaf sensitivity

Group I Very sensitive	Group II Sensitive	Group III Less sensitive
Hazelnut (*Corylus avellana*)	European beech (*Fagus silvatica*)	Red oak (*Quercus rubra*)
Riesling grape (*Vitis vinifera*, Riesling)	Rowan tree (*Sorbus intermedia*)	English oak (*Quercus pedunculata*)
Birch (*Betula verrucosa*)	Gingko (*Gingko biloba*)	Hedge maple (*Acer campestre*)
Apple (*Malus communis*, Golden Delicious)	Black locust (*Robinia pseudoacacia*)	Blue spruce (*Picea pungens*)
Poplar (*Populus* sp.)	Horse chestnut (*Aesculus hippocastanum*)	Mugho pine (*Pinus mugo*)
Norway spruce (*Picea abies*)	Norway maple (*Acer platanoides*)	Lawson's cypress (*Chamaecyparis lawsoniana*)
Nordman's fir (*Abies nordmanniana*)	Nikko fir (*Abies homolepis*)	Juniper (*Juniperus communis*)
White pine (*Pinus strobus*)	Scotch pine (*Pinus silvestris*)	
European larch (*Larix europaea*)	Austrian pine (*Pinus nigra austriaca*)	
Japanese larch (*Larix leptolepis*)		

Table 19. HCl resistance of agricultural and garden plants based on leaf sensitivity

Group I Very sensitive	Group II Sensitive	Group III Less sensitive
Yellow lupine (*Lupinus luteus*)	Oats (*Avena sativa*)	Kale (*Brassica oleracea acephala*)
Red clover (*Trifolium pratense sativum*)	Winter rye (*Secale cereale*)	Tobacco (*Nicotiana tabacum*, Bel W 3)
Broad bean (*Vicia faba*)	Winter barley (*Hordeum vulgare*)	Beet (*Beta vulgaris esculenta rubra*)
Bush bean (*Phaseolus vulgaris nanus*)	Tomato (*Lycopersicum esculentum*)	Gynura (*Gynura aurantiaca*)
Radish (*Rhaphanus sativus vulgaris*)	Black currant (*Ribes nigrum*)	Fatshedera (*Fatshedera lizei*)
Lettuce (*Lactuca sativa capitata*)	Gooseberry (*Ribes grossularia*)	Rosebay rhododendron (*Rhododendron catawbiense*)
Strawberry (*Fragaria chiloensis ananassa*)	Iris (*Iris germanica*)	Fuchsia (*Fuchsia hybrida*)
Barberry (*Berberis vulgaris*)	Primula (*Primula melacoides*)	Lily of the valley (*Convallaria majalis*)
	Tuberous begonia (*Begonia tuberhybrida*)	Lilac (*Syringa vulgaris*)

Fig. 28. Vegetation-free zone near a heavy metal smelter in Sudbury, Ontario. (Photo: H. Schönbeck)

tion gradient near a single pollutant source. When emissions are very high or when conditions for distribution of the pollutant are poor, all vegetation can be totally destroyed. Around a source that emitted about 6000 t of SO_2 daily (Fig. 28), vegetation-free areas were found up to a distance of 30 km. Figure 29 shows vegetation bordering on this polluted area.

To describe the zonal injury resulting from a pollutant gradient, the area around an iron ore smelter that had a daily SO_2 output of about 10 t has been selected as an example (Fig. 30). Emission conditions are given in Table 1. The slightly diluted waste gas impinged directly on the exposed slopes of the valley in which the smelter was located. All vegetation was killed up to a radius of 500 m from the source (Guderian and Stratmann, 1962, 1968; van Haut and Stratmann, 1970). In general, vegetation on projecting hilltops, gulleys or in cross valleys was most severely injured.

The change from the severely eroded denuded zone to the transition zone beyond, is characterized by the occurrence of clusters of resistant ground plants, such as *Erica cinerea*, *Galium mollugo*, *Veronica officinalis*, *Dechampsia flexuosa*, *Rumex acetosa*, and *Convallaria majalis*, which is very sensitive to HF. The first shrubs to be found were *Rhamnus frangula* and *Sambucus racemosa*. As in all ecotone areas (Lötschert, 1969), interspecific competition played no role here.

In the grass and scrub-wood zone, clumps of hardy grasses, such as hair grass (*D. flexuosa*), begin to form connected plots, and finally a fairly continuous cover. Under protection of the grass, oak shoots (*Quercus petraea*) are able to survive. Wherever the oak grows above the grass, the impact of sulfur dioxide on the

Fig. 29. Vegetation island in a protected valley *(above)* and transition zone 30 km from a heavy metal smelter *(below)*. (Photo: H. Schönbeck)

exposed side can be seen in the sparse discolored foliage and from the typical "streamlined" growth habit, which Halbwachs and Kisser (1967) have also observed on spruce and birch exposed to HF.

With increased distance from the source, the oak and red beech shoots are able to survive and form thickets. Species diversity of the plant community increased with a decrease in pollutant concentration. At a distance of about 1.5–2 km from the source, the composition of the beech-oak *(Fago-Quercetum)* forest typical for the acid soils in the area was found again.

In plant-ecologic studies near an aluminum smelter Nikfeld (1967) found that, close to the pollutant source, the typical vegetation community *(Epilobio-Seneci-*

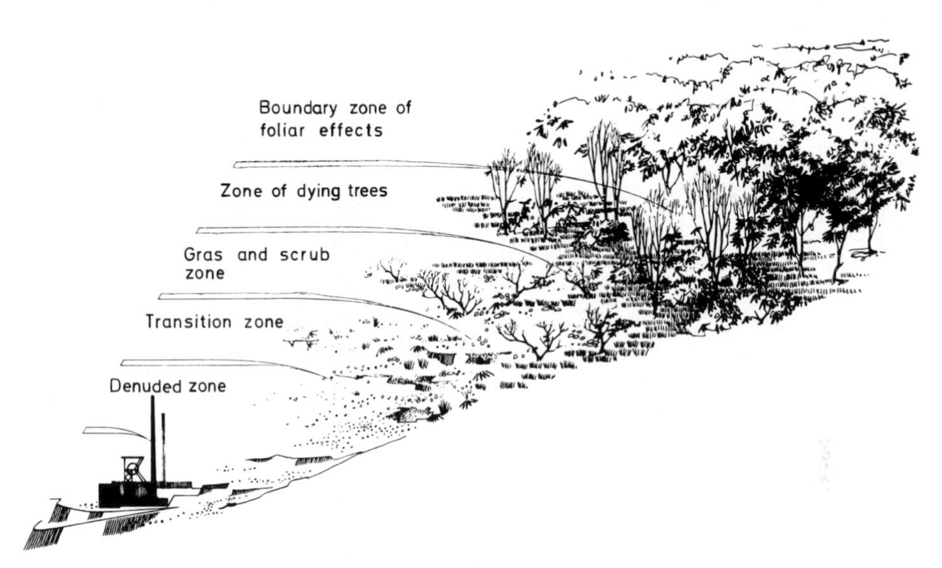

Fig. 30. Schematic drawing showing zones of injury on an exposed slope near an SO_2 source. (After van Haut and Stratmann, 1970)

onetum silvatici) changed to a fluoride resistant community consisting mainly of composites *(Tanaceto-Artemisietum)*.

Brandt and Rhoades (1972), in an area with heavy pollution from limestone dust, showed that *Quercus prinus* and *Acer rubrum* were replaced by *Liriodendron tulipifera* and *Quercus alba* as dominant tree species in the forest community. A distinct decrease in importance of *Quercus rubra* was also observed.

Changes in composition of plant communities cannot be explained solely through direct pollutant effects on sensitive species. In studies on pure and mixed cultures, Guderian (1966b) showed that pollutants also caused changes in interspecific competition (Fig. 31). The primary effect on sensitive species increased so that they are at a disadvantage in competition for necessary growth factors such as nutrients, water, radiation, and growing space. Finally, these species lose importance in composition of the community (Fig. 32, see p. 127).

As a result of changes in competition, the reduction of the sensitive members of the community usually results in better growth of resistant species, when pollutant stress is not too extreme. In this case, the total yield of the culture is not as greatly reduced as would be expected from disappearance of sensitive species.

Such changes reduce the value of a crop as fodder. A reduction in legumes causes a reduction in the amount of protein and minerals and an increase in raw fiber (Klapp, 1953). The soil enrichment effects from legumes through nitrogen fixation, nutrient decomposition, and maintenance of friability are also lost (Klapp, 1967).

Just how important the sensitivity of papillionaceous plants is for their natural distrubiton in areas polluted with SO_2 is unknown. It is also not clear what effect atmospheric pollution has on the agricultural cultivation of clover varieties (Bochow, 1965).

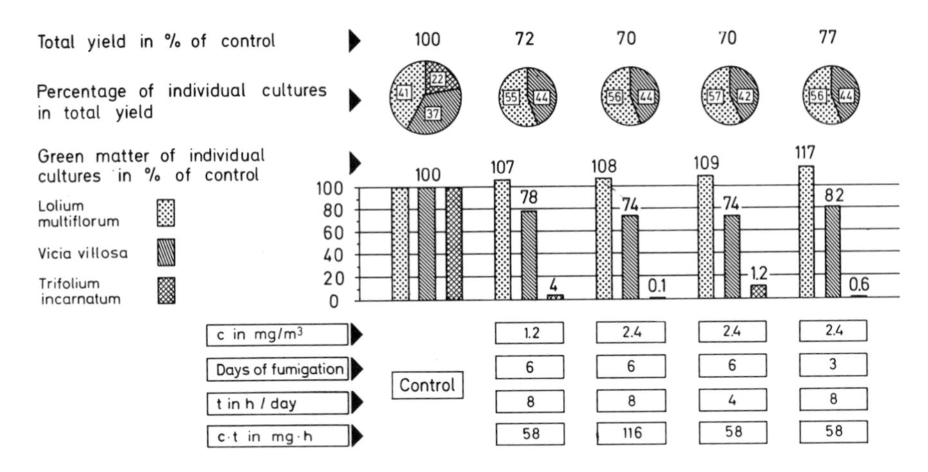

Fig. 31. Changes in yield and structure of a community consisting of rye grass, hairy vetch, and crimson clover after exposure to SO_2

The most important methods for reducing effects of air pollutants at the growing site can be derived from genetic- and environment-determined resistance characteristics of single plant species and varieties. If these characteristics are to be used optimally in practice for recommendations for cultivation and for diagnostic purposes, effects-criteria and concentration of the given pollutant must also be considered in the determination and evaluation of the degree of resistance.

The importance of effects-criteria for the evaluation of plant resistance is shown in the following example. Based on foliar necrosis, *Lupinus luteus* is the most sensitive of various legumes, followed by *V. faba* and *V. sativa*. If resistance is based on green matter production, however, *Lupinus* is much more resistant than the other two species (Guderian, 1966b). Certain deciduous trees show leaf necrosis before conifers, but continue to grow in areas where conifers have died off (Wentzel, 1968).

The intended use of the plant is a criterion for the evaluation of effects when selecting species and varieties suitable for cultivation in polluted areas. With one criterion, for example resistance groups based on leaf sensitivity (O'Gara, 1922; Thomas and Hendricks, 1956; Thomas, 1961), essential information for diagnostic purposes can be obtained, but this is not enough for the selection of species suitable for a polluted area.

Recommendation of unsuitable species can occur in practice when use is not considered. In the American literature, certain conifers are classified on the basis of foliar sensitivity in the resistant group, while annuals were classified as very sensitive (Thomas, 1961). No differentiation was made between the resistance of the assimilatory organs as a part of the plant and the resistance and capacity for regeneration of the total organism (Vogl et al., 1965). In fact, the listed annuals can be successfully cultivated in areas where conifers can no longer survive (Wentzel, 1968). An evaluation of resistance should be based on the extent of reduced productivity.

Larch serves as a typical example to show changes in resistance characteristics with pollutant concentration (Guderian and Stratmann, 1962; Wentzel, 1963). Under SO_2 concentrations high enough to cause acute injury, necrosis to needles occurs on larch—*L. europaea*, as well as *L. leptolepis*—before spruce *(P. abies)* and pine *(P. silvestris)*. Under long-term exposure to low concentrations, however, larch is one of the most resistant conifers (Wentzel, 1963) and is frequently used for recultivating pine and spruce forests destroyed by pollutant action.

The genetically determined differences in resistance, especially of individuals, lead to possibilities for selecting and breeding relatively resistant varieties (Zieger, 1953, 1954; Pelz and Materna, 1964). After Krüger (1951) promoted such breeding work, selection and vegetative reproduction of resistant spruce and pine was begun about a decade ago (Rohmeder et al., 1962; Rohmeder and von Schönborn, 1965). Bell and Clough (1973) found reductions in yield of about 50% in the S23 cultivar of *Lolium perenne*, after exposure to 0.191 mg SO_2/m^3 air for six months. A native variety from a polluted area showed no injury. It is assumed that the long-term exposure of the native population to SO_2 leads to a natural selection of resistant lines. Therefore, the chances for breeding of a resistant cultivar appear to be good.

3. Comparisons of the Phytotoxic Characteristics of Sulfur Dioxide, Hydrogen Fluoride, and Hydrogen Chloride

Aside from the degree of phytotoxicity, effects of a particular air pollutant depend on uptake, distribution, mechanism of action and excretion by plant organs. The extent of accumulation, compared with the variation in natural content of the particular element, determines the diagnostic value of chemical analysis for proof of pollutant effects. The possibility of latent reductions in productivity of agronomic cultures and natural ecosystems under the influence of long-term low level concentrations is also discussed.

3.1 Functions in Plant Metabolism

The elements sulfur, fluorine, and chlorine, because of their universal distribution, are found in all plant species. They are present, however, in different amounts which vary with habitat conditions and stage of development of the plant (Thomas et al., 1950b; Arnold, 1955; Thomas and Alther, 1966; Guderian, 1970). These elements also have different functions in plant metabolism.

3.1.1 Sulfur

Sulfur, as an essential element, occurs in the plant in organic and inorganic forms. Of all plant organs, the leaves have the highest sulfur content. Organically bound sulfur varies between 0.06% in conifer needles and 0.7% in leaves of certain Crucifera. Neutral sulfur is found in organic substances as sulfhydryl, disulfide, and sulfonic groups or in heterocyclic rings. In the SH form, sulfur is important in production of the essential amino acids cysteine and methionine. Through oxidation, the sulfhydryl compounds are converted to sulfide compounds. The redox system of cysteine/cystine and tripeptide glutathione determines the formation of the hyper structure of amino acids. The activity of several enzymes is also dependent on the high reactivity of the SH group, for example coenzyme A. A physiologically important disulfide bridge occurs in liponic acid and is active in the breakdown of pyroracemic acid and in the biosynthesis of acetyl CoA.

It also plays a part in the Calvin cycle of photosynthesis (Bassham et al., 1962). Sulfonic esters occur mostly in glycosides and their physiologic importance is not yet entirely clear (Mengel, 1968). Sulfur forms a heterocyclic bond in the coenzymes thiamine pyrophosphate (Vitamin B_1) and biotine (Vitamin H).

Sulfur that is not bound organically occurs in the plant as sulfate and, depending on amounts taken up from the soil and the air, can greatly exceed organic sulfur content. Therefore, total sulfur content varies more than neutral sulfur content (Guderian, 1970). In metabolism, sulfate activates fermentation, maintains colloidal structure of protoplasm, increases assimilation activity, and has a more positive effect on carbohydrate formation than chloride (Burghardt, 1962). Translocation of assimilates is not hindered by sulfur and the activity of hydrolyzing enzymes, which has been reduced by excessive amounts of chloride, can be stimulated by addition of sulfur (Latzko, 1954). Because of its role in metabolism, it is clear that sulfur is a plant nutrient that cannot be replaced by any other element. Excessive sulfur is stored primarily in cell vacuoles as sulfate (Marschner and Michael, 1960) and serves as a sulfur reserve. Content of organically bound sulfur increases very little or not at all (Thomas et al., 1943, 1944b; Schmalfuß, 1964; Guderian, 1970). In extreme cases, sulfate content can increase to five or ten times of the organic sulfur content before injury occurs. Injury from excessive soil sulfur occurs only under extreme habitat conditions on soils with high amounts of calcium sulfate or through over fertilizing (Penningsfeld and Forchthammer, 1965) and is of little importance in practice. Outside of industrialized areas, for example in the USA, Canada, New Zealand, Australia, and East Africa, large areas can be found where sulfur deficiency occurs, as reported in the reviews of the literature by Freney et al. (1962) and Koronowski (1969).

Higher plants take up sulfur through the roots mainly as sulfate. An additional supply of sulfur comes from assimilation of SO_2 in the air. Cotton plants were shown to be able to take up to 50% of their total sulfur from the air (Olsen, 1957). Fumigation of sulfur-deficient sunflower, corn, and tobacco plants with SO_2 for several weeks, showed that SO_2 in the air can be a decisive source for sulfur and serves to promote normal plant growth (Faller, 1970). Under 1.5 mg SO_2/m^3 air, enough to cause necrosis, the amount of sulfur taken from the air amounted to between 86 and 93% of total sulfur (Faller, 1968). Sulfur taken up from the air was found in the same fractions as that from the soil (Thomas et al., 1943; Harrison et al., 1944; Faller and Höfner, 1968) but proved to be less effective physiologically (Thomas et al., 1944b).

The described effects where SO_2 has served as a source for nutrient sulfur were shown on sulfur-deficient plants. A positive effect from uptake of SO_2 from the atmosphere can only be expected on soils where availability of sulfur is low. Based on experimental results of Leone and Brennan (1972), plants on such soils should be able to tolerate higher SO_2 concentrations (see Sects. 2.5 and 2.2). Saalbach et al. (1962), in experiments with various soils from Westfalia, found slight increases in yield of marrowstem kale *(B. oleracea* var. *medullosa)* with soils supplied with ammonium sulfate nitrate when compared with those supplied with nitrate of lime and ammonium. That does not mean, however, that sulfur deficiency occurs over large areas of middle and western Europe, as can also be seen from sulfur budgets and air analyses from large areas. On land used for agricultural purposes in the Federal Republic of Germany in the 1950s, about 13 kg sulfur/ha was removed from soils through harvested products (Kurmies, 1957), about the same amount as phosphorus. Losses due to washout were estimated to be about 60 kg/ha. This yearly loss of 73 kg/ha is replaced by about 19 kg from fertilizer and about 68 kg through precipitation. Based on the assumption that a representative amount of sulfur from total SO_2 emissions enters the soil through precipitation, Ulrich (1972) estimates yearly input of sulfur at about 100 kg sulfur/ha. Values from Kurmies (1957) indicate a yearly excess of sulfur of 14 kg and those from Ulrich (1972) an excess of 25 kg/ha. These calculations assume an equal distribution of sulfur over the entire Federal Republic and do not take into account the particular requirements of individual crops. Precipitation in polluted areas can have a much higher sulfur content than in nonindustrialized areas and this can lead to an increase in soil acidification (Jordan and Bardsley, 1959; Garber, 1967). According to Thomas (1958), the yearly addition of sulfur from precipitation varies between 5 and 250 lbs/acre, depending on the amount of SO_2. Sulfates can be relatively easily washed out and, especially on light, permeable soils, an accumulation of sulfates in the ground water cannot be avoided (Kick and Kretzschmar, 1968).

Even in areas far from industry, substantial amounts of sulfur enter the soils through certain land-use practices. In an input–output analysis carried out in a spruce stand in Solling, Ulrich (1972) showed that with 150 kg, additional sulfur was three times that of a beech stand or of nonwooded areas. With needles and the large amount of dead wood, a spruce tree has a very large surface area. Especially after precipitation, large amounts of SO_2 are absorbed in the surface water films, causing a substantial filter effect. Through-fall water, with a pH of about 3, causes washout of Ca and Mg in leaf litter and leads to acidification of the upper soil strata. In mineral soils, the hydrogen ions from sulfuric acid can cause a release of aluminum ions from clay minerals, which often has negative effects in surface water and fish ponds. According to Norwegian studies, the commonly observed fish kills of Scandinavia have been traced back to SO_2 in air masses that come from England and the Continent during certain high pressure conditions (Ottar, 1972). In waters of the primitive rock areas in southern Sweden, conductivity has increased by about 20% and the sulfate content has increased by about 50 μmol. Bicarbonate content was reduced (Malmer, 1972). Air analyses in southern Norway show SO_2 values that are several times higher than in the northern part of the country, where values of 2 μg SO_2/m^3 air have been measured (DFG, 1964).

3.1.2 Fluorine

According to present knowledge fluorine, present as fluorides, is not an essential element for normal plant development (Mitchell and Edman, 1945; Thomas and Alther, 1966) as opposed to sulfur and chlorine. As demonstrated in hydroponic cultures, however, fluoride can have a stimulatory effect (Baumeister and Burghardt, 1957). As with the other two elements, fluorides can be taken up from soil and air; leaves have the highest levels. Natural fluoride levels, that is, fluoride content in plants grown in areas free of industry, are low, usually less than 20 ppm in dry matter (Bredemann, 1956; Thomas and Alther, 1966; Garber, 1967). Exceptions include certain species of the Theaceae. For example, Wang et al. (1949) found fluoride levels in tea plants of between 57 and 355 ppm in dry matter, with the highest values in the oldest leaves. According to Bredemann (1956), fluoride content varies greatly among cultivars. Samples of tea from the Congo showed levels between 5.8 and 6.0 ppm in dry matter, while samples from China varied between 37.5 and 390 ppm in dry matter. McLendon and Gershon-Cohen (1955) found fluoride levels in Camellia plants fertilized with superphosphate of between 121 and 1370 ppm in dry matter. Aside from these "accumulator plants," fluoride content of the soil has little or no effect on fluoride levels in vegetation. After uptake of fluoride from the nutrient medium, levels in the roots are always higher than in above-ground plant parts. This relationship is reversed after exposure to atmospheric HF (Brennan et al., 1950; Garber et al., 1967).

In certain plants, such as *Dichapetalum* spp. in South Africa and *Acacia georginae* Gidyea in Australia (Peters and Shorthouse, 1964; Peters et al., 1965), formation of fluoroacetic acid (FCH_2COOH) takes place and severe losses through death of pasture animals can occur. This relatively nonpoisonous compound is converted to fluocitric acid through enzyme action in the body of the animal and inhibits the citric acid reaction in the Krebs cycle (Peters et al., 1953). The potential poisonous activity of fluoroacetic acid can be seen from LD_{50} values, for example 1 mg/kg live weight for horses and 0.35 mg/kg for Guinea pigs.

Since fluoride does not play a role in the metabolism of most plants, a detoxification of fluoride in the plant cell does not take place. Jacobson et al. (1966) found that fluoride almost always occurs in the ionic form. Another reason for the high toxicity of fluoride may also be that it accumulates to high levels in the chloroplasts, as Chang and Thompson (1965) have shown on isolated cell organelles. An increase in fluoride effects is dependent on secondary translocation in the leaf (see Sect. 2.5.4).

Hazards to vegetation come almost solely from direct action of gaseous fluoride compounds on above-ground plant parts. Indirect effects from fluoride accumulation in soils has only been observed close to heavy emittors (Garber et al., 1967). Only a small portion of fluoride in the soil is available to plants. This also explains the low natural levels of fluoride in plants, as fluoride, with 0.07%, is about as common in the earth's crust as phosphorous and sulfur (Teworte, 1971). Similar to zinc and lead (Guderian and Schönbeck, 1971), large amounts of fluoride in soils can lead to long-term effects to vegetation over and above injury from atmospheric fluoride. In lysimeter studies, MacIntire et al. (1958) have shown that fluoride, as opposed to sulfur and especially to chloride, is only slightly washed out of soils.

Another difference between effects of fluoride and those of sulfur and chloride is that, after accumulation of fluoride in fodder plants, injury to animals can occur. Extensive economic losses to agriculture occur from fluorosis, especially of cattle, which is presently the most important disease directly traceable to air pollutant effects (Rosenberger, 1963; Rosenberger and Gründer, 1968). The symptom-complex of chronic fluorosis, caused by long-term uptake of fodder contaminated with low levels of fluoride, include defects in the enamel of incisors, osteophytes and extoses of long hollow bones, ribs, and lower jaw bones, as well as disruption of enzyme functions (Schmid, 1956; Suttie, 1969). Difficulties in eating, lack of appetite, lameness, loss of weight, reduction of fertility, and a reduction of milk and meat production can result (see Sect. 4.1.2).

3.1.3 Chlorine

As with sulfur and fluorides, chlorine, present in the chloride form, can be taken up either through roots or through leaves. The consensus of investigators states that chloride uptake from the nutrient substrate is directly proportional to the supply, as presented in the literature reviews of Arnold (1955), Burghardt (1962), and Leh (1969). Uptake by plants, which favor uptake of chloride ions over other anions, can be explained from the fact that, after NO_3^-, Cl^- comes before SO_4^{2-} and PO_4^{3-} in the lyotropic ion series. Chloride ions with their small volume and high uptake rates repress the uptake of other anions. Cations combined with chloride ions are more favored in uptake than, for example, when combined with sulfate ions (Geissler, 1953). The relationship between inorganic cations and anions in the plant is shifted strongly toward the latter through chloride.

Chloride levels in the plant are determined primarily through supply of chloride ions. There are essential species-specific differences, but chloride content could never be useful as a taxonomic characteristic. The average portion of osmotic values due to chloride is about 20% for cultivated plants, weeds, and other wild plants with a natural variation between 0.5 and 60% depending on habitat. Halophytes have an average chloride content of about 65% with extreme values up to 95% (Arnold, 1955).

Physiologically, chloride is involved in many different metabolic reactions. Hydrophilic chloride ions, in contrast to sulfate ions, have an imbibing effect on colloidal protoplasma, a fact which leads to wilt resistance, induced through chloride (cf. Fig. 13). Increases in yield of chlorophilic plants may come about through replacement of nitrate ions with chloride ions in colloidal chemical reactions (Schmalfuß, 1950).

Aside from effects on protoplasm, chloride ions also alter effects of enzymes. Latzko (1954) found an absolute reduction in the activity of hydrolyzing enzymes of plants supplied with different chloride-containing fertilizers. This reduced enzyme activity, and the strong hydration of the chloride ions can lead to disruption of the carbohydrate budget and protein synthesis, which may affect yield and quality of the plants (Burghardt, 1962). A great excess of chloride can lead to destruction of chlorophyll through photo-oxidation (Boresch, 1939) resulting in necrosis. Lesser excesses lead to a reduction of the amount of chlorophyll, which is expressed as chlorosis. After addition of chloride, Schulze (1957) found that the relationships of chlorophyll a/chlorophyll b and carotine/xanthophyll remained the same, but slight changes between the green and yellow pigments occurred. Pirson (1958) also concluded that an excess of chloride can directly inhibit photosynthesis. Observations of stimulatory effects contrary to this allow no conclusions to be made regarding the functional relation between chloride content and photosynthesis (Pirson, 1958). Later studies by Arnon (1959) showed that chloride activated electron transport between the cytochromes during photophosphorylation.

As is generally known, chloride promotes growth of various plants, especially those of the Chenopodiaceae (Buchner, 1951). This and other effects of chloride, however, are fairly nonspecific. Studies by Ulrich and Ohki (1956) and Johnson et al. (1957) indicate that chloride is a plant nutrient in the sense defined by Kick (1969). The requirements for chloride, in spite of its participation in various important metabolic reactions, are in the range of the micronutrients (Stout and Johnson, 1957) and are fully covered through natural chloride levels in the air and from precipitation. Chloride deficiency, as opposed to sulfur deficiency, is not known to have occurred. Of practical importance is only an excess of chloride, which is expressed in the chlorosis, necrosis, and inhibition of growth already mentioned, as well as in succulence and xeromorphy. It is apparently unimportant in this case whether chloride is taken up from the nutrient substrate through the roots or from the air through the leaves (van Haut and Guderian, 1976; see also Sect. 3.2).

Distribution of chloride taken up through the roots varies according to growth conditions and plant species, but chloride levels usually increase from the root toward the leaves. Chloride levels in leaf veins and petioles are much higher than in intercostal tissues (Arnold, 1955). Accumulation of chloride in leaf margins and tips does not necessarily occur in all plants, as shown in the example in Section 2.5.4. Chloride levels in the vascular fluid are usually quite low (Walter and Steiner, 1936) but may be quite high in the membranes of the leaf veins. Like fluoride, but unlike sulfur, chloride is not covalently bound in metabolism. That many plants with a large percentage of chloride in the dry matter continue to grow normally indicates that chloride has a low phytotoxicity in comparison with fluoride and sulfur.

In the soil, chloride is present almost exclusively in a dissolved form, so that washing out is common. According to extensive lysimeter studies by Pfaff (1958), the chloride content of European soils is fairly constant. Supply and demand remain in balance through uptake from soil and air by plants, as well as washout and replacement through precipitation and fertilizers. Arnold (1955) reported, however, that chloride levels in nonfertilized areas with a mild climate, such as in Western Europe, were approaching a minimum. Indirect injury to vegetation via the soil by an accumulation of chloride from air pollutants is not to be expected. Effects on soil structure and pH, for example through conversion of calcium in easily soluble compounds, will probably only occur near large emittors.

Chloride levels in the air not resulting from human activity vary with the distance from oceans. In Sweden, values of 0.1 to 30 μg Cl/m^3 air were found (Egner, 1956) and in Germany and Austria, values between 2.6 and 54.5 μg Cl/m^3 air were reported (Riehm and Quellmalz, 1959). Yearly addition of chloride through precipitation in various parts of the world varies between 5.1 and 447 kg/ha (Eriksson, 1952).

3.2 Accumulation of Pollutants in Plant Organs

The question of the accumulation of pollutants in plant organs is discussed according to the following points:

1. How high is the accumulation of S, Cl, and F compared with levels of these elements in the air?
2. What are the relationships between pollutant accumulation and effect of the pollutant?
3. What value do chemical analyses of plants for S, Cl, and F have for showing the presence of these elements as air pollutants?

3.2.1 Accumulation of S, F, and Cl in Connection with Pollutant Dosage

Depending on plant species and on concentration, HF is 10 to 1000 times more harmful than SO_2. The phytotoxicity of HCl is comparable to that of SO_2 (see Sect. 4.1). These reactions result not only from differences in toxicity, but are also dependent on uptake of the single pollutants, as shown in Figure 33 and in Table 20.

Figure 33 presents an experiment in which winter rye (*S. cereale*, F. von Lochows Petkuser) and winter barley (*H. vulgare*, Peragis 12 Melior) were fumigated during tillering with low concentrations of SO_2, HCl, and HF. Concentrations were chosen so that no necrosis would occur. The experiments were carried out simultaneously in the field to assure comparable growing conditions.

After fumigation with 0.46 mg SO_2/m^3 air for 421 h, the sulfur content of winter rye increased by 0.15% and of winter barley by 0.28% in the dry matter. Fumigation with 2.31 μg F/m^3 air for the same time caused an accumulation of fluoride of 155.8 and 139.6 ppm in dry matter in rye and barley respectively and exposure to 0.31 mg HCl/m^3 air for 319 h caused an increase in chloride of 0.77% in rye and of 0.73% in barley.

Table 20 presents values for a determination of the connection between pollutant accumulation and pollutant supply. Values are calculated as the quotient of

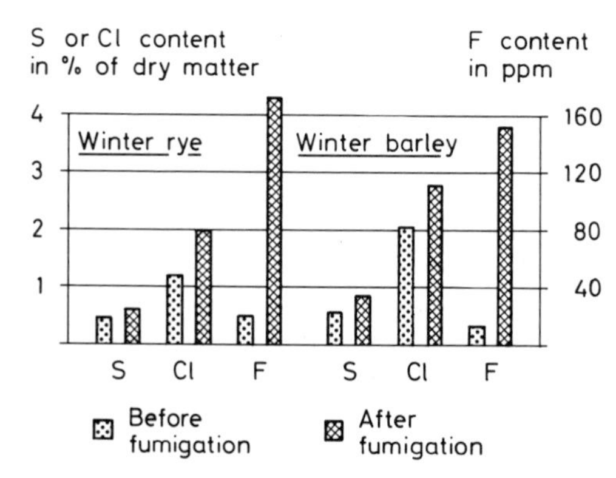

Fig. 33. Sulfur, chloride, and fluoride content of winter rye and winter barley before and after exposure to SO_2, HCl, and HF

Table 20. Quotients of pollutant accumulation and pollutant supply compared with accumulation of S, Cl, and F

Plant species	S accum./ S supply	Cl accum./ Cl supply	F accum./ F supply	Relative Cl accum. (S accum. = 100%	Relative F accum. (S accum. = 100%)
Secale cereale	1.3	8.1	13.6	620	1050
Hordeum vulgare	2.3	8.6	12.0	370	520
Pinus ponderosa	0.51	0.69	4.0	135	784
Picea abies	0.63	0.68	4.3	130	680
Fragraria chiloensis	1.2	1.3	16.5	110	1370
Chrysanthemum indicum	4.0	5.18	29.8	130	740
Average	1.66	4.1	13.4	249	857

the pollutant accumulation and pollutant dosage, which is taken as the product of concentration of the particular element and exposure time. An average of 1.66 for SO_2 compares with 4.1 for HCl and 13.4 for HF. If the quotient sulfur accumulation/sulfur dosage is equal to 100, then the percentage of chloride equals 249 and the percentage of fluoride equals 857, as an average for the six plant species. This means that chloride was accumulated 2.5 and fluoride 8.5 times more than sulfur.

One can only speculate upon causes for this phenomenon, which is also a determining factor for the extent of injury from SO_2, HCl, and HF. One important factor is certainly the different rates of absorption in the water film in the cutinized stomatal chambers.

Uptake of fluoride through the epidermis could also be important. This is not enough, however, to explain the great differences between the three pollutants, as experiments with hypostomatic leaves of *Fatshedera lizei* have shown (see Sect. 2.5.3.2). It may be that fluoride, from HF in the air, enters the leaf not only through stomata but also directly through the cuticle and epidermis, but not in amounts large enough to explain the differences described above. The permeability of the epidermis to fluoride salts has already been shown by Brewer et al. (1960b). This ability of leaves to absorb nutrients and other substances directly through leaf tissues has been used to advantage in leaf fertilizers, pesticides, and herbicides (Wittwer and Bukovac, 1969). Ectodesma, which are often difficult to detect, may play a major role in the transport of substances from the leaf surface to the mesophyll (Franke, 1964, 1967; Tukey et al., 1962).

3.2.2 Relationship Between Accumulation of Pollutants and Plant Injury

Sulfur dioxide is a gaseous pollutant that clearly causes either acute or chronic injury, depending on its concentration in the air. With HF, the difference between these two types of injury is quite gradual and the action of HCl lies between SO_2 and HF.

Acute necrotic injury from SO_2 is caused only by short-term exposures to high concentrations. Studies on sulfur accumulation after exposure to a constant dosage $(c \cdot t)$ consisting of different concentrations and exposure times have shown that, after short exposures to concentrations high enough to cause acute injury, SO_2 uptake is relatively low, due to a blockage of gas exchange shortly before death of leaf tissue. In this case it is not the amount of absorbed sulfur, but rather the rate of absorption, which is the deciding factor. The action of high concentrations on fast-growing plants under high light and humidity conditions and a favorable temperature can cause acute necrosis at levels of absorbed sulfur that are difficult to detect through chemical analysis of the plant tissue. This does not mean, however, that ambient SO_2 concentrations, which cause acute injury, cannot be detected through chemical analysis of leaves, as was assumed from earlier fumigation experiments with very high concentrations (Guderian, 1970). In polluted areas, high concentrations do not occur alone but in combination with low concentrations. Because of the comparatively high contribution of low concentrations to the total frequency of occurrence of SO_2, chemical analysis of leaves for the detection of SO_2 is less problematic the higher the pollutant load. Leaves with the least necrosis should be taken for analysis, however.

Through the action of low concentrations, SO_2 is either oxidized, reduced, or forms organic compounds at the same rate as it is taken up (Thomas et al., 1944a, b). Under such long-term low-level exposures, sulfur content can increase to two or even 2.5 times normal before chlorosis occurs.

Sulfur content in plant organs, therefore, can only be used as an indicator for SO_2 activity, but not as an effects criterion for SO_2 (Guderian, 1970). Because of pollutant controls, based especially on the air quality principle (Stratmann, 1972), SO_2 concentrations that cause acute injury occur quite seldom. Long-term stress of low concentrations has increased, however, as a result of an increase in total emissions and building of taller stacks for better dispersion of pollutant gases. As opposed to pollutant conditions of a few decades ago, the major hazard to vegetation today is the chronic exposure to low SO_2 concentrations, which occur over very large areas.

Necrosis, as a result of short-term exposure to high HF concentrations, does not often occur. Fluoride concentrations that could cause such injury are much higher than concentrations that cause extensive chronic injury through long-term exposures. Such conditions usually occur only in areas near heavy emittors or during unfavorable weather conditions. In areas with fluoride pollution, however, necrotic injury is widespread, but as a result of long-term accumulatory effects in the leaves rather than as a result of short exposures to high concentration peaks. During this accumulation phase, chlorosis usually occurs before lethal levels are reached. There is usually a chlorotic zone dividing the necrotic zone from healthy leaf tissues. The presence of both injury types on the same leaf occurs more commonly with HF than with SO_2 (Guderian et al., 1969; van Haut and Stratmann, 1970; Jacobson and Hill, 1970). The occurrence of leaf injury indicated a minimal toxic accumulation that is at least several times higher than the normal fluoride content. There is, therefore, a relatively close relationship between fluoride content of the leaf and the occurrence of necrosis. Fluoride levels can then be used as an indicator of effects and may be considered as a decisive effects criterion for the establishment of air quality criteria (Hill, 1969). Limitation of fluoride levels in fodder is the easiest and safest means of preventing fluoride toxicity in animals (see Sect. 4.1).

Comparative studies on the accumulation of fluoride from the atmosphere and injury to vegetation have led to grouping of plants according to their resistance to fluoride (Brandt, 1971). Very sensitive species, such as many members of the Iridaceae and Liliaceae and various *Prunus* and *Pinus* species (Bolay and Bovay, 1965), show chlorosis or necrosis of leaves with fluoride levels under 50 ppm in the dry matter. Sensitive plants, which show injury at levels of 50–200 ppm include the rest of Iridaceae and Liliaceae, other conifers and several members of the Papilionaceae and Gramineae. Relatively resistant plants show visible injury at levels over 200 ppm. Through analysis of fluoride levels in plants, an evaluation of the risk to vegetation can be made, although difficult situations under highly variable ambient concentrations cannot be overlooked. Action of high concentrations can cause injury before the level normally considered to be injurious is reached.

Fluoride levels in the plant alone yield little information as to how high concentrations in the air may be before presenting a hazard to vegetation and animals. For determining potential hazards and setting allowable pollutant levels, studies of the quantitative relationships between fluoride level and effects on plants are necessary. This may be possible through comparison of the fluoride content of standardized indicator plants (Scholl, 1971a) with the reactions of other species. Such relationships provide a basis for setting and monitoring allowable pollutant levels in the air.

Available information on chloride allows no detailed evaluation of the connection between chloride accumulation and effects. In general, injury occurs after uptake of large amounts of chloride, which can be detected through chemical analysis of the plant material. It has also been observed that accumulation of chloride from air pollutants has a more severe effect than that taken up from the soil. It is possible that uptake of chloride from the air leads more quickly to changes in anion/cation relations, as chloride in the nutrient medium promotes uptake of cations (Geissler, 1953).

3.2.3 Pollutant Content in Plant Organs as an Indicator for Pollutant Activity

Knowledge of the extent of pollutant accumulation and uptake rate is an important basis for the determination of hazards to vegetation, as shown in Sect. 3.2.2. The diagnostic value of chemical analyses for determining the presence of pollutant injury is based on the relationship between the amount of the accumulated element and the variation of the normal content of that element. Optimal conditions for this sort of diagnosis are plant responses in the form of necrosis or chlorosis which occur after uptake of relatively large amounts of the particular pollutant. The normal content of the given element must be comparatively low.

Figure 34 presents a comparison of levels of S, Cl, and F in spruce needles and leaves of red beech exposed to these pollutants (right) until the occurrence of chlorosis or necrosis with normal levels (left) of these elements. The greatest variation in normal levels is with chloride, which had a relationship of 1:10 between the lowest and highest values. Fertilization with substances containing chloride would cause much higher levels. Variation of normal levels of fluoride, and especially of sulfur is quite small.

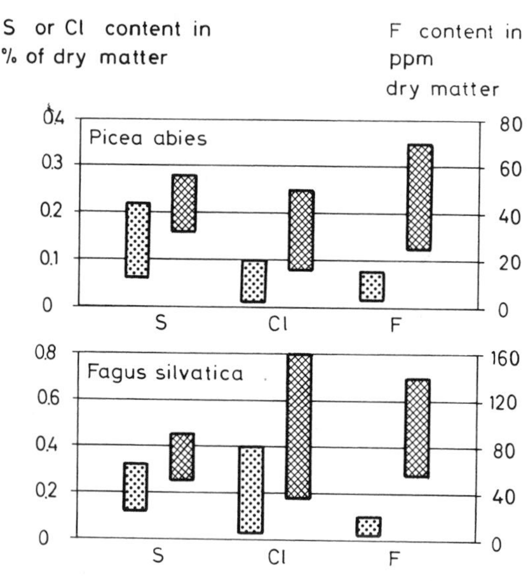

Fig. 34. Extent of pollutant accumulation up to appearance of leaf injury compared with the natural variation in S, Cl, and F content

Compared with normal levels, the lowest accumulation occurs with sulfur. Accumulation to two or three times the normal level assumes a relatively long exposure to high SO_2 concentrations (Guderian, 1970). Uptake of chloride is greater than that of sulfur—see Table 20, in Sect. 3.2.1—but the variation of the normal levels is also greater. Because of the comparatively low accumulation of sulfur and chloride, levels of these elements cannot be used as an effects-criterion, but only as an indication of pollutant action.

Chlorosis or necrosis on plants with low normal levels of sulfur and chloride can occur at accumulated levels that are close to the average normal levels for these elements, because of the extreme variability of normal levels. Single leaf samples, therefore, should not be used for deciding whether pollutants containing sulfur or chloride have been taken up. Chemical analysis for these elements should only be used as a comparative method. Effects of these pollutants can be determined only after a comparison of several samples from the study area.

Accumulation of fluoride from atmospheric pollutants, however, can usually be detected without such comparison samples. High accumulation from exposure to pollutants and the low normal levels of fluoride allow the use of chemical analyses as an absolute method for determining pollutant action (Guderian and van Haut, 1970), as is shown in Figure 34. Spruce and red beech are both fairly sensitive to fluoride and visible injury occurs at relatively low levels of accumulation. Levels at which foliar injury occurs, however, are generally higher than the range of variability of the normal levels. In resistant species, injury occurs at levels which are very much higher than the normal fluoride content. With an increase in the extent of accumulation, conditions become better for determining the action of fluoride through chemical analysis of the plant material.

3.3 Hidden Injury

The present air pollution situation is characterized by persistent effects of low ambient concentrations and the simultaneous occurrence of several pollutants. General predictions as to the risk of injury from air pollutants are difficult to make, especially since there are few data available from relevant long-term investigations. Under such circumstances, it is difficult to show evidence of the effects of air pollutants as, in many cases, there are no obvious external symptoms of injury (Guderian and van Haut, 1970). Since air pollution affects large areas, the question of latent injury has become an acute problem, as is indicated by the revival of interest in the debate concerning "hidden injury" (Hill et al., 1958; McCune et al., 1967; Keller and Schwager, 1971).

Table 21, with reference to work from Vogl et al. (1965) and Weinstein and McCune (1971), presents type and extent of effects of air pollutants on various levels of organization of the ecosystem as a basis for further discussion. In this way, effects at the cellular level are brought into relationship with reactions of the entire plant or even the entire plant community. The effects-process begins with uptake of the pollutant into the cell. It can be inferred that, for example, the inhibition of enzymes by fluoride can lead to disruption of metabolism in vivo, as has been shown in vitro through work on enolase, phosphoglucomutase, and succinate dehydrogenase (Slater and Bonner, 1952; Weinstein and McCune, 1971). In areas where ambient air contains fluoride, Keller and Schwager (1971) have shown that such reactions occur under natural conditions. Peroxidase activity was increased and tissue aging was accelerated with an increase in fluoride concentration. Such reactions can occur without visible symptoms of injury (Table 22). Changes in natural metabolic reactions, as well as changes in organelles and cell compartmentalization, lead finally to injury or death of the cell.

Changes in the cell which exceed a particular intensity also affect the next highest level of organization. Structural changes in the lamellar system of chloroplasts, as shown in the EM photographs in Figures 10 and 11, for example, lead to reduction in CO_2-assimilation and hinderance in growth and development, especially of leaves; cell injury may be expressed externally as chlorosis or necrosis.

The effects described for single organs can lead to increased sensitivity of the plant to climatic factors and pathogens, or directly to reductions in growth and quality. These changes, as well as changes in competition between species, cause transformations and reduction of species diversity in the vegetation cover. Under extreme pollution conditions, areas devoid of vegetation may occur.

Accumulation of pollutants in fodder plants to concentrations dangerous for man and animals is further evidence for the far-reaching effects of air pollution on ecosystems, as is a disruption of biogeochemical cycles and a reduction in capability of the ecosystem for self-regulation.

Therefore, it is evident from Table 21 that, before symptoms of injury can be seen, many different reactions occur at the cell and organ levels, which can be classified as "hidden injury". This term, first used in 1899 by Sorauer and Ramann, has been repeatedly enlarged upon and modified (Haselhoff and Lindau, 1903; Stoklasa, 1923; Reckendorfer, 1952; Kisser, 1966) but is still basically defined as functional injury to a plant through air pollutant action without visible

Table 21. Classification of air pollution effects on plants.

Cell	Level of Organization		
	Tissue or Organ	Organism	Ecosystem
Pollutant uptake	Pollutant uptake or deposition	Pollutant uptake or deposition	Pollutant accumulation in the plant and other ecosystem compartments such as soil and surface and ground water
Alteration in cell milieu	Altered assimilation, respiration or transpiration	Growth modification	
Effects on enzymes and metabolites			
Modification of cell organelles and metabolism	Changes in growth and development	Increased susceptibility to biotic and abiotic influences	Damage to consumers as a result of pollutant accumulation (i.e., fluorosis)
		Reduction in production, yield, and quality	Changes in species diversity also due to shifts in competition
Disruption of reaction pathways	Chlorosis Necrosis	Death of plant	Disruption of biogeochemical cycles
Cellular modification	Death or abscission of plant organs		Disruption of stability and reduction in the ability for self-regulation
Disruption and death of cell			Breakdown of stands and associations
			Expansion of denuded zones (desolation)

indications of this action. Many attempts have been made to verify this logically based postulate, but only recent work has proved the existence of this type of injury (Koritz and Went, 1953; Hull and Went, 1952; Guderian et al., 1969). These investigations, as well as a comparison of the effects mechanisms, indicate that greater hidden injury is to be expected through action of HF than from SO_2. HCl can also cause reductions in yield without visible morphologic changes. Experiments on red clover (*T. pratense* var. *sativum*, Red Head, tetraploid), which were carried out in fumigation chambers at the field station described previously, provide an example. As shown in Table 22 plants fumigated with 0.12 mg HCl/m^3 air for 50 h exhibited a significant reduction in growth but were completely free of visible signs of injury. Radish *(R. sativus* var. *radicula)* also showed a significant reduction in yield, although there were no visible symptoms on above-ground plant organs (van Haut and Guderian, 1976).

There are, however, no convincing arguments for retaining the term hidden injury. In agreement with McCune et al. (1967) it must be stated that air pollution effects on vegetation constitute a scientific problem and a problem for practical pollution control. For questions of a scientific nature, such a term is irrelevant.

Table 22. Yield and degree of injury of *Trifolium pratense* after exposure to various HCl concentrations for 50 h

Treatment	Yield in g dry matter/pot[a]	Degree of injury
Control	136 ± 5.89	—
50 h · 0.12 mg HCl/m^3	115 ± 7.07[b]	—
50 h · 0.20 mg HCl/m^3	102 ± 4.88[b]	0.5

[a] Average of five pots.
[b] Significant for 99.9% of control.

The decisive questions for intensive agriculture, as well as for natural ecosystems, however, is to what extent does economic or ecologic damage occur through encroachment upon intended use (see Sect. 2.1). The major difficulty is the identification of effects in practical situations and the confirmation of cause and, especially, magnitude of injury.

Effects that are difficult to recognize without visible symptoms of injury include not only reduction of growth and changes in predisposition, but also inferior productivity when chlorosis or early senescence occurs. Brewer et al. (1960 b) found significant reductions in growth and foliation on orange trees exposed to HF, although only slight chlorosis of the leaves was visible. Annuals exposed to SO_2 (Guderian, 1970) and HF (Hitchcock et al., 1963) also showed a significant depression in yield, whereas other symptoms of injury were only slightly visible. After exposure of *Lolium perenne* cultivar S23 to 0.191 mg SO_2/m^3 air for six months, Bell and Clough (1973) found a reduction in growth of 52% compared with control plants. This resulted from reduced tillering, lowered production of new foliage, and early senescence of leaves. It is doubtful if such responses would be recognized as coming from the action of air pollutants without control experiments in filtered air, as Bleasdale (1952) had already shown in experiments with the same species of grass in filtered and nonfiltered greenhouses in an industrial area. These examples clearly show that a lessening of the expected value of vegetation can occur, either without any visible symptoms at all or with symptoms that would be recognized only with difficulty as coming from exposure to air pollutants.

The increment-boring method, further developed by Vins (1965, 1971), makes possible the control of production processes in forestry. This point has been more or less ignored for agronomic or horticultural plants. As long as air pollutants, in concentrations that cause acute injury, only occurred in the close vicinity of the source, there was also no real need. The readily visible leaf necrosis which gradually increased toward the source, guaranteed recognition of pollutant action and evaluation of the extent of injury. Such obvious responses do not occur, however, under long-term exposures to the low concentrations present in ever increasing areas. The possibility of hidden injury is great in such cases. Pollutant-free times during which plants could recover, do not often occur and long-term action of low concentrations is effective within a broader spectrum of ecologic factors than short-term action of high concentrations.

As can be seen from results of experimental investigations, depression of yield over large areas occurs not only in forestry, but also in fruit culture and agronomic cultivation. Reduction of yield and growth, as shown in Table 21, are the end results of a chain reaction that began with uptake of the pollutant into the plant cell. All previous attempts to use a particular response of cells or organs to determine the extent of growth depression have yielded unsatisfactory results (Guderian and Stratmann, 1968; Hill, 1969). Such single responses are important indications of pollutant activity, but are usually not useful as a criterion for evaluating the effect (Guderian, 1970). The actual extent of disruption of production processes in the various branches of agronomic land use and in ecosystem studies usually can only be determined through growth and vegetation analyses. Consideration must be made of the particular method of exposure (see Sect. 4.2). To solve such economic, as well as ecologic, problems, monitoring networks, comparable with those used for air analyses in large areas (Prinz and Ixfeld, 1971; Buck and Ixfeld, 1971), are necessary, as suggested by Ellenberg (1973) and others. Experiments in which indicator plants were exposed to various pollutant concentrations over large areas (Prinz and Scholl, 1975; Scholl and Schönbeck, 1975) indicate that such methods, along with records of pollutant concentrations at single monitoring sites, can be used in determining the extent of the hazard from air pollutants for wild and cultivated plants, as well as for humans.

4. Discussion of the Suitability of Plant Responses as a Basis for Air Pollution Control Measures

In the development of measures for pollutant control and the selection of methods for their realization, the following points must be considered: high pollutant concentrations, which cause acute injury, usually occur only during production problems or accidents at the pollutant source, or during unfavorable weather conditions. As a result of increasing density of industrial and domestic areas and traffic, the pollution situation is characterized by persistent effects of low ambient concentrations. Plants have little or no chance for detoxifying absorbed pollutants, due to the lack of sufficient pollution-free times and as a result of accumulatory effects in the plant organs and soil. The number of biologically active pollutants that occur together is increasing. Because of effects of pollutants occurring in combination or alternately, increasing risks of injury to vegetation must be expected which, at the present time, cannot be evaluated. If optimal results for the protection of vegetation are to be achieved, investigations of pollutant effects must concentrate on such conditions.

The suitability of plants, as a basis for air pollution control measures rests on the extreme sensitivity of higher and lower plants and on specific plant responses to single pollutants. Results of studies on effects of pollutants serve not only directly for establishment of protective measures for vegetation, but also as a basis for determining air pollutant control measures. Effects to vegetation, therefore, are important criteria in setting allowable air pollutant limits, in the use of plants as biologic indicators of air quality, and in the establishment of control measures for reducing pollutant stress.

4.1 Air Quality Criteria and Their Use in Setting Air Quality Standards

Air quality criteria are the quantitative relationships between the degree of pollution of the air with a particular component or mixture of components and reactions of humans, animals, plants, or materials (Prindle, 1966; Prinz and Stratmann, 1969). They give an indication of the type and extent of effects of a particular pollutant, defined by concentration and duration of action, on the group of objects being studied. They are medical or scientific indications related to effects that have been determined through scientific studies or from practical experience (VDI, 1974). Air quality criteria do not take technical or economic feasability into account. These dose–response relationships, also considered as injury limits, play an important role in practical pollution control based on the principle of air quality management (Persson, 1971; Stratmann, 1972). They serve as the scientific basis for the legislature in the setting of standards in the form of maximum ambient pollution concentrations (Ta-Luft, 1974).

4.1.1 Air Quality Criteria

A calculated risk is included in the legally established standards (BImSchG, 1974). Therefore, they must be based on representative quantitative relationships between pollutant and effect. Such relationships exist when pollutant load and effects are characterized and evaluated with the aspects presented in Sections 2.1 to 2.3 and when effects are studied on plants with a normal or approximately normal predisposition (see Sect. 2.5).

Only a small part of all experiments dealing with pollutant effects on vegetation fill these requirements. Because enough information on pollutant stress is sometimes not available, a large number of observations and descriptions of effects to agronomic land-use cannot be considered (Kaudy et al., 1955; Hölte, 1960; Bolay and Bovay, 1965; Dässler and Grumbach, 1967; Bolay et al., 1971a).

Effects criteria that do not allow an evaluation of effects on the useful value of the plant, for example certain biochemic reactions, cannot be used directly. Experiments in which plants show abnormally weak responses, such as under conditions of low light intensity (Rohmeder et al., 1962) or low air exchange rates in fumigation chambers (Zattler and Chrometzka, 1964), are also unusable.

Under consideration of these postulates, the following selection and evaluation of experimental results in regard to their suitability for the general evaluation of possible hazards to vegetation is given. Experiments with unrealistically high concentrations, that is concentrations which seldom occur even in heavily polluted areas, are not included. It is important to notice that only the harmfulness of a particular concentration can be determined, since the effects of injury approach infinity.

4.1.1.1 Sulfur Dioxide

Threshold concentrations have been determined over a period of several years for a large variety of important agricultural plants in field experiments near an SO_2 source (Guderian and Stratmann, 1962, 1968; Stratmann 1963b). Table 23 gives the average SO_2 concentrations—compare with Section 1.1.1—$\bar{c}_i$ for the exposure time t_i and $\bar{c}_m$ for monitoring time t_m. At the lower threshold concentrations no injury was observed, but at the higher concentrations there were significant effects on growth, yield, or quality of the particular plant. For example, apple (Ellison's orange) at station V, with a $\bar{c}_i$ value of 0.22 ppm and a $\bar{c}_m$ value of 0.010 ppm was not injured. At the next station (IV) closer to the source, with values of 0.26 ppm and 0.020, however, yield and quality of fruit were reduced.

Table 23. Threshold values for SO_2 from a field study near an SO_2 source

Plant species	Threshold concentration in ppm	
	$\bar{c}_i$ during exposure time t_i	$\bar{c}_m$ during monitor time t_m
Fruit trees and berry shrubs		
Gooseberry *(Ribes grossularia)*	0.22–0.26	0.010–0.020
Currant *(Ribes rubrum)*	0.22–0.26	0.010–0.020
Apple *(Malus communis)*	0.22–0.26	0.010–0.020
Sour cherry *(Prunus cerasus)*	0.22–0.44	0.010–0.083
Sweet cherry *(Prunus avium)*	0.26–0.44	0.020–0.083
Prune *(Prunus domestica)*	0.26–0.44	0.020–0.083
Forest cultures		
Spruce *(Picea abies)*	0.22–0.26	0.010–0.020
Pine *(Pinus silvestris)*	0.22–0.26	0.010–0.020
Larch *(Larix europea)*	0.22–0.26	0.010–0.020
Red beech *(Fagus silvatica)*	0.22–0.26	0.010–0.020
English oak *(Quercus pedunculata)*	0.22–0.26	0.010–0.020
Grains and summer rape		
Winter wheat *(Triticum sativum)*	0.24–0.28	0.009–0.024
Winter rye *(Secale cereale)*	0.28–0.31	0.024–0.051
Summer wheat *(Triticum sativum)*	0.23–0.38	0.015–0.050
Oats *(Avena sativa)*	0.23–0.38	0.015–0.050
Summer rape *(Brassica napus)*	0.23–0.57	0.015–0.124
Truck crops		
Potato *(Solanum tuberosum)*	0.21–0.23	0.010–0.015
Red beet *(Beta vulgaris)*	0.28–0.31	0.024–0.051
Fodder plants		
Red clover *(Trifolium pratense)*	0.28–0.31	0.024–0.051
Alfalfa *(Medicago sativa)*	0.28–0.31	0.024–0.051
Oats (for green fodder)	0.23–0.38	0.015–0.050
Winter rye (for green fodder)	0.28–0.31	0.024–0.051
Summer rape (for green fodder)	0.28–0.31	0.024–0.051
Vegetables		
Spinach *(Spinacia oleracea)*	0.22–0.25	0.010–0.020
Carrot *(Daucus carota)*	0.28–0.49	0.024–0.104
Tomato *(Lycopersicum esculentum)*	0.31–0.57	0.051–0.124

In Figure 35, the effects on stem and lateral growth of red beech *(F. silvatica)* are shown. Up to station IV, 1350 m from the source with a $\bar{c}_i$ value of 0.26 ppm and a $\bar{c}_m$ value of 0.020 ppm, both growth processes have been significantly reduced after exposure for three years. Increase in basal area, as a measure for lateral growth, clearly shows this reduction. With the same annual ring width, basal area (G_z) increases linearly with the diameter at the start of the experiment (D) and quadratic with increase in diameter (D_z=double annual ring width), as

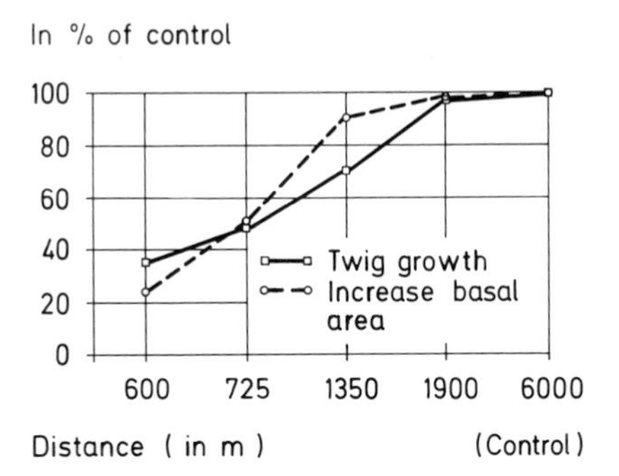

Fig. 35. Reduction of growth and basal area of red beech approaching an SO_2 source

shown in the following formula:

$$G_z = \frac{\pi}{4}(D+D_z)^2 - \frac{\pi}{4}D^2 = \frac{\pi}{4}(2D \cdot D_z + D_z^2).$$

In a long-term field study in Sudbury, Ontario, Dreisinger (1970) and Linzon (1970) found that average concentrations of 0.005 to 0.01 ppm SO_2 caused injury to white pine *(P. strobus)* and aspen *(P. tremula)*, as well as to sensitive agronomic crops, such as buckwheat *(F. esculentum)*, barley, and red clover. Near an iron smelter in Trail, British Columbia, injury to *Pinus ponderosa* occurred at a concentration of about 0.03 ppm SO_2, with occasional peaks up to 0.5 ppm (Dean and Swain, 1939). The very low averages in the studies described above result from the large number of pollution-free time periods during the total study time, as is common near single sources (see Sect. 2.3). Short-term action of relatively high concentration peaks is mainly responsible for the observed injury.

Materna (1965b, 1966) and Materna et al. (1969) report on the responses of *Picea abies* in the Erz mountains of Czechoslovakia after long-term exposure to low levels of SO_2. Injury still occurred at average concentrations of 0.05 mg during the monitoring time and 0.15 mg/m^3 air during the exposure time. Peak concentrations for sampling intervals of 30 min, rarely exceeded 0.5 mg SO_2/m^3 air. Recent studies by Materna (1972, cited in Knabe, 1973) show that slight injury to spruce stands *(P. abies)* in northern Bohemia occurs at average concentrations of 0.025 mg SO_2/m^3 air. The possibility of a combined effect with fluoride, chloride, or nitrogen oxides could be excluded (Table 24). In the Ruhr area, older—severely injured—pine stands *(P. silvestris)* can only be found in areas where the average concentration does not exceed 0.18 mg SO_2/m^3 air (Knabe, 1970a).

The field studies described here were carried out under considerably different pollutant conditions: one was characterized by high peak concentrations and long pollutant-free times, the other by long-term action of low concentrations. Based on the average concentrations, however, the injury limits are in the same range. The extreme phytotoxicity of the pollutant type with high peak concentrations results from the progressive increase of acute injury with an increase in concentration, whereas injury from the pollutant type with long-term action of low concentrations rests on cumulative effects of the pollutant and the absence of

Table 24. Degree of injury to spruce stands in northern Bohemia compared with SO_2 concentration and sulfur content of needles. (After Materna, 1972; cited in Knabe, 1973)

	Level of injury	Average SO_2 concentrations			
		Entire year μg/m³	During growing period μg/m³	During winter μg/m³	Average sulfur content of needles (%)
0.	No injury observed	15	5	30	0.1
I.	Slight injury, slight reduction in assimilation	25–35	10–20	50–60	0.135
II.	Medium injury, significant reduction in assimilation, isolated cases of death of trees	30–40	20–30	50–70	0.165
III.	Severe injury, significant reduction in assimilation, opening of stand through death of trees	50–70	20–50	70–90	0.240
IV.	Death of trees in all age groups	70–90	40–70	100–200	0.320

pollutant-free times which allow the plant to recover (see Sect. 2.2 and 2.3). This is especially true for long-term effects on fruit and forest cultures, as slight effects can add up over a period of years or even decades and cause a reduction of the useful value of the plant.

Injury limits derived from fumigation experiments under conditions similar to natural conditions, tend to be slightly higher than those derived from field studies, usually because of the shorter exposure times. Values established in early work (Wislicenus and Neger, 1914) have not been considered because of the inadequate analytical and technical methods of the time. Thomas and Hill (1937) found no reduction in photosynthesis or respiration of alfalfa *(M. sativa)* under an SO_2 concentration below 0.3 ppm, except after long-term exposures where chronic effects from sulfate accumulation were observed. Scheffer and Hedgcock (1955) report injury to *P. ponderosa* and *Larix occidentalis* after exposure to 0.5 ppm SO_2 for 7 h, Katz and McCallum (1952) found weak symptoms on larch needles after an 8 h fumigation with 0.3 ppm. In the author's experiments with European larch *(L. europaea)*, the first necrosis of needles was observed after exposure to 0.45 mg SO_2/m^3 air for 2 h. In continuous fumigations over 150 h Zahn (1961) found that foliar injury occurred to alfalfa *(M. sativa)*, crimson clover *(T. incarnatum)*, and hairy vetch *(V. villosa)* at concentrations above 0.15 ppm and to summer wheat *(T. sativum)*, oats *(A. sativa)*, and spinach *(S. oleracea)* at concentrations above 0.20 ppm. Bell and Clough (1973), after exposure to 0.191 mg SO_2/m^3 air for 6 months, found that growth of *Lolium perenne*, breeding clone "Aberystwyth S23" was reduced by 52%. This resulted largely from reduced tillering and foliation, and accelerated aging of leaves. It is possible, therefore, that reductions in yield, found in earlier experiments (Bleasdale, 1952) with S23 in greenhouse chambers with filtered and nonfiltered industrial air, resulted from SO_2 concentrations, which did not exceed 0.1 ppm.

4.1.1.2 Hydrogen Fluoride

Among the fluoride containing air pollutants, the gaseous, water-soluble compounds have the highest phytotoxicity. Hydrogen fluoride is of the most practical importance. It is emitted from many sources and occurs through breakdown of other fluoride compounds, such as F_2, SiF_4, and H_2SiF_6. Since these gases have a comparable phytotoxicity (Hitchcock et al., 1963), studies on the effects on vegetation are concentrated on hydrogen fluoride.

Gaseous fluoride compounds are also indirectly, that is after accumulation in plant material, responsible for air pollution injury to animals (Oelschläger, 1971). Fluoride uptake through inhalation, even in areas with severe fluoride pollution, has little effect on animals, and particulate fluorides that come from the soil or from the air as sedimented dusts and that are adsorbed on fodder plants, are also relatively harmless, because of their low solubility and digestability (Oelschläger, 1970; Freitag et al., 1970).

For these reasons, the evaluation of hazards to vegetation and animals is based on the determination of gaseous fluoride compounds in the atmosphere (McCune et al., 1965; Oelschläger, 1971; Brandt, 1971). However, the collection of a certain amount of the particulate fluorides by a sampling method assumed to be selective for gaseous fluorides cannot be avoided (Weinstein and Mandl, 1971; Buck and Stratmann, 1965). Depending on the selectivity of the sampling method, fluoride levels that cause injury in the field would be too high, because of particulates. In the presence of only gaseous fluoride compounds, the value would be correspondingly lower.

For evaluation of the risks to plants and animals, two criteria should be considered: the levels of gaseous fluoride compounds in the air, and fluoride accumulation in plants. The veterinary aspect is fairly straightforward (Brandt, 1971). For example, in order to avoid chronic fluorosis with economically negative results in milk cows, average fluoride content in fodder of 30 to 40 ppm *F* in dry matter should not be exceeded (Shupe et al., 1962; Schmidt et al., 1968; Shupe, 1969; Suttie, 1969; NAS, 1971).

Extensive studies on the relationship between fluoride accumulation in the plant and atmospheric fluoride concentrations have been carried out. The often-cited dose–rate relation:

$$F = KCT$$

where: F = increase in F content above the normal level
C = concentration of gaseous fluoride compounds in the atmosphere
T = exposure time

indicates the effort that has been given to quantify this relationship (Brandt, 1971). The coefficient, K, represents the various internal and external growth factors that have an important influence on pollutant uptake. Such a simplified dose–rate relation can be of great use when one considers that K is a variable and that only approximations can be expected. Comparative studies of the accumulation of atmospheric fluoride and effects on plants have led to the formulation of resistance-groups (Brandt, 1971), as discussed in Section 3.2.2.

The evaluation of risks to vegetation under strongly variable fluoride concentrations poses a particular problem, because the effect is not only dependent on the accumulated amount of fluoride but also on the uptake rate. Fluoride content in plant organs allows an evaluation of the danger to animals and, with limitations, to plants, but this is not sufficient for the setting of allowable limits for atmospheric fluoride at which injury to plants and animals does not occur. For this purpose, studies on the quantitative relationship between atmospheric fluoride and effects on vegetation are necessary.

In a field study near a fluoride source, gladiolus varieties and Ponderosa pine showed necrosis after exposure for two weeks on sites with average concentrations of 0.77 and 0.49 ppb HF (Adams et al., 1956, 1957). After continuous exposure to 1–5 ppb HF for two years, orange trees showed an inhibition in stem and lateral growth, reduced leaf size with severe chlorosis, retardation in blossom formation, lower yields, and a reduction in external quality and vitamin C content of fruits (Brewer et al., 1960a). Similar effects were found after exposure to 1.0 ppb HF for 27 months (Thomas and Alther, 1966).

In fumigation experiments with sensitive forest trees, such as spruce *(P. abies)*, white pine *(P. strobus)*, Nordman's fir *(A. nordmanniana)*, Rowan tree *(Sorbus intermedia)*, and red beech *(F. silvatica)*, such extensive leaf injury occurred after exposure to 1.3 µg HF/m^3 air for several days that long-term exposures would probably cause severe reduction of growth (Guderian et al., 1969). Severe effects on quality were found on sensitive ornamentals, such as tulips, daffodils, hyazinths, and crocus, after exposure to average concentrations of 1.5 and 2.0 µg HF/m^3 air for 12 days. Injury occurred on gladiolus and begonia *(B. tuberhybrida)* after exposure to 1 µg HF/m^3 air for only three days. Growth of winter barley was strongly reduced after exposure to 3.3 µg HF/m^3 air for 290 h at the time of tillering.

With an average concentration of 0.85 µg HF/m^3 air, possibly toxic fluoride accumulation up to 85 ppm F in dry matter was found in certain grass and clover species. At the slightly higher concentration of 1.1 µg HF/m^3 air, fluoride accumulation up to 330 ppm was found. Growth was not influenced. Red and white clover *(T. pratense* and *T. repens)* and alfalfa *(M. sativa)* showed slight necrosis and chlorosis, while Welsh and German rye grass *(L. multiflorum* and *L. perenne)*, orchard grass *(Dactylis glomerata)*, meadow fescue *(Festuca pratensis)*, and timothy *(Phleum pratense)* remained free of injury.

Benedict et al. (1964) found similar effects. For example, fluoride accumulation in alfalfa reached 124 ppm F in dry matter with no effect on growth after exposure to 0.8 µg HF/m^3 air for 5 weeks. A significant reduction in yield of alfalfa, orchard grass, and lettuce *(L. sativa* var. *romana)* was found after exposure to 100 µg HF/m^3 air for 20 days. Red clover and timothy, from a clover grass mixture, had fluoride accumulation of 51 ppm in dry matter after exposure to 1.9 µg HF/m^3 air for 7 days (MacLean et al., 1969). A fluoride content of 78 ppm was found after exposure of alfalfa to 5 µg HF/m^3 air for 40 h (Benedict et al., 1965).

After fumigation of sorghum for eight days at 2.55 ppb HF, Hitchcock et al. (1963) found that flowering was inhibited; after 15 days at this concentration significant chlorosis and necrosis was observed and yield was reduced by 33%. In another experiment, carried out during the elongation and anthesis stage of development at 0.82 ppb HF, significant reduction of growth occurred after 14 days. In fumigation experiments at 1.5 ppb, 5 ppb, and 10 ppb, Adams et al. (1957) observed leaf injury on sensitive plants, such as peach *(P. persica)*, gladiolus, larch *(L. occidentalis)*, lilac *(Syringa vulgaris)*, and Douglas fir *(Pseudotsuga menziesii)* after exposure for 65–110 h at the lowest concentration. *M. sativa*, as the only fodder plant in this study of 40 different species or varieties, showed an F accumulation of 149 ppm F in dry matter after exposure to 1.5 ppb HF for 240 h. Spierings et al. (1971) found severe tip burn on the tulip variety Paris after exposure to 0.6 μg HF/m^3 air for 16 days. The very sensitive gladiolus variety Picardy was severely injured after exposure to only 0.1 ppb HF for five weeks (Compton and Remmert, 1960). Microscopic analysis of needles from *Pinus ponderosa*, which was exposed for 40 h to 0.5 ppb HF, showed granulation of protoplasm, increase in the number of vacuoles, and disintegration of mesophyll cells (Solberg et al., 1955).

Figure 36 shows a summary of data from experiments with various HF concentrations (McCune, 1969). It can be seen that the most sensitive plants, such as conifers, gladiolus, and sorghum, are injured at a concentration of about 0.5 μg HF/m^3 air; certain species or varieties are injured below this concentration. Through the slow rise of the curve, as compared with SO_2, the slight influence of concentration on degree of injury becomes evident. Through long-term exposure, very low atmospheric fluoride concentrations can cause extensive injury (see Sect. 2.2).

4.1.1.3 Hydrogen Chloride

Little information on the effects of HCl on vegetation is available. In fumigation experiments in small greenhouses under conditions similar to natural conditions (see Sect. 1.1.2), sensitive deciduous trees and shrubs showed significant

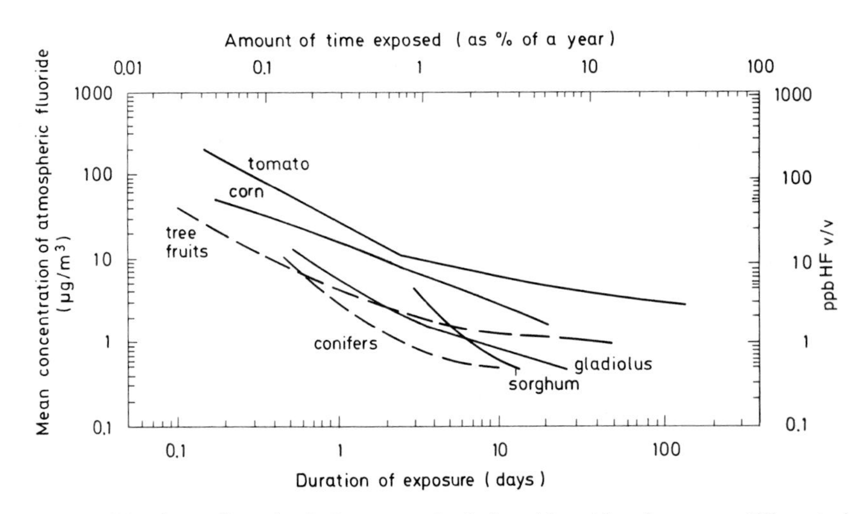

Fig. 36. Possible air-quality criteria for atmospheric fluoride, with reference to different plant species. (Reprinted with permission from McCune, 1969)

Table 25. HCl concentrations that cause injury to vegetation

Plant species	Average concentration in mg/m^3 air	Exposure time in h	Type and extent of injury
Vitis vinifera	0.08	90	slight necrosis slight chlorosis
Grape	0.28	114	severe necrosis
Sorbus intermedia	0.08	90	slight necrosis
Rowan tree	0.28	114	medium necrosis
Aesculus hippocastanum	0.08	90	slight necrosis
Horsechestnut	0.28	114	medium necrosis
Populus sp.	0.08	90	slight necrosis
Poplar sp.	0.28	114	medium necrosis
Fagus silvatica	0.08	90	very slight necrosis
European beech	0.28	114	medium necrosis
Robinia pseudoacacia	0.08	90	very slight necrosis
Black locust	0.28	114	slight necrosis
Corylus avellana	0.12	140	slight necrosis
Hazelnut	0.19	150	medium necrosis
Betula verrucosa	0.11	229	slight necrosis
Cut-leaved birch	0.18	228	medium necrosis
Tilia platyphyllos	0.11	229	slight necrosis
Basswood	0.18	228	medium necrosis
Acer campestre	0.11	229	very slight necrosis slight chlorosis
Hedge maple	0.18	228	slight necrosis
Gingko biloba	0.11	229	no injury
Ginko	0.18	228	slight necrosis
Quercus pedunculata	0.11	229	no injury
English oak	0.18	228	very slight necrosis
Pinus strobus	0.08	90	no injury
White pine	0.28	114	slight necrosis

foliar injury at an average concentration of only 0.08 mg HCl/m^3 air after 90–230 h. This suggests that a reduction in growth and yield would occur after longterm exposure (van Haut and Guderian, 1976). The extreme phytotoxicity of HCl is evident when one considers that exposure was discontinuous for 6 h per day (Table 25).

After exposure for 360 h on 34 days, a significant reduction in yield occurred on broad beans *(V. faba)* at 0.09, on lupine *(L. luteus)* and spinach at 0.17, and on winter rye *(S. cereale)* at 0.24 mg HCl/m^3 air. After intermittent exposure to 0.12 mg HCl/m^3 for 47 h, red clover *(T. pratense)* showed a statistically significant reduction in growth without visible symptoms of injury (see Table 22 and Sect. 3.3). Effects of low HCl concentrations on radish are particularly significant.

Radish yield was reduced by 20%, compared to the control, with only a slight reduction in leaf weight after exposure to 0.12 mg HCl/m^3 air for 140 h. Yield of

Welsh rye grass *(L. multiflorum)* was reduced by 10% under the same conditions. As shown in electron-microscopic studies, action of 0.13 mg HCl/m^3 air for 43 h led to an acceleration of the aging processes of spinach (see Sect. 2.2). Ewert and Dässler (1969) observed foliar injury on such plants as lettuce *(L. sativa)*, pea *(P. sativum)*, hornbeam *(Carpinus betulus)*, red alder *(A. glutinosa)*, and spruce *(P. abies)* after continuous exposure to 0.2 mg HCl/m^3 air. Intermittent exposure to 0.05 mg HCl/m^3 air for 300 h caused no visible injury even on sensitive plants. After continuous exposures for 120–200 h, flecking of leaves of sensitive plants occurred which, after 350–400 h developed into necrosis. Relative resistant conifers, such as *Picea omorica*, *Pinus silvestris*, and *Pinus nigra*, had no visible injury after exposure to 0.05 mg HCl/m^3 air for 460 h.

In summary, it can be said, that intermittent exposure to 0.05 mg HCl/m^3 air presents only a slight risk of injury to sensitive plants. It is probable, however, that injury will occur in areas with constant HCl concentrations, because of accumulation effects. From fumigation experiments, it appears that free chlorine (Cl_2) is more phytotoxic than HCl (Brennan et al., 1965). Alfalfa and radish proved to be the most sensitive plants and leaf necrosis occurred after exposure to 0.10 ppm for only 2 h. Exposure for 4 h to 0.10–0.25 ppm caused injury to tobacco, zinnia *(Zinnia elegans)* corn, and sunflower. The most resistant plants proved to be petunia *(Petunia hybrida)* bush bean *(P. vulgaris)*, geranium *(Pelargonium zonale)* and azalea, which exhibit injury after exposure to 0.8 to 1.0 ppm for 4 h.

4.1.2 Air Pollution Standards

Two basic principles can be recognized in an analysis of the methods used in various countries for air pollution control (TA Luft, 1974; Stratmann et al., 1968, Persson, 1971). In the concept of best practical means, the avoidance of pollutant effects is determined by economic or technical practices. There is, however, no relation to air quality. In the air quality principle, the air should be kept as clean as required and air pollution control is based on the prevention of certain air pollution effects.

The setting of such air pollution standards requires a knowledge of the quantitative relationship between the pollutant and the object to be protected. Of these objects, such as humans, animals, plants, or materials, plants show especially sensitive responses to the pollutants under consideration. Dose–response relationships determined for plants provide a basis for setting standards for SO_2, HF, and HCl.

It must be mentioned, however, that standards are not always identical with the dimensionless dose–response relations (VDI, 1974; Buck, 1970). More often, economic and social aspects are considered in the establishment of standards. Standards, therefore, are a compromise between the scientifically determined necessity for air quality and technical and economic possibilities for the reduction of emissions. In the following, the standards in West Germany for the protection of vegetation from the action of SO_2, HF, and HCl are presented with a discussion of allowable pollution limits in other countries (see Sect. 4.1).

The following standards for SO_2 are presented in the Technical Guidelines for Air Pollution Control (TA Luft, 1974):

$$IW_1 = 0.14 \text{ mg } SO_2/m^3 \text{ air – long-term value,}$$
$$IW_2 = 0.40 \text{ mg } SO_2/m^3 \text{ air – short-term value.}$$

For the monitoring of pollution limits for gaseous air pollutants under practical conditions, the values I_1 and I_2 have been derived:

$$I_1 = \bar{x}$$
$$I_2 = \bar{x} + ts$$

where: $\bar{x}$ = average of single values for a particular study area

$$s = +\sqrt{\frac{2(\bar{x} - x_1)^2}{2z - 1}}$$

x_1 = single values greater than $\bar{x}$
z = number of single values greater than $\bar{x}$
t = factor of a one-tailed Students t test, where $t = 1.64$ at the 95% confidence level.

Through observance of values set in the TA Luft ($IW_1 = 0.14$, $IW_2 = 0.40$) most plants can be protected from injury due to SO_2. In agriculture, it is possible that acute injury will still occur on spinach and walnut (*Juglans regia*) and in forestry, injury may occur especially on spruce and fir. A greater risk of injury to certain epiphytes, such as foliose lichens, is also present, but protection, in general, has been greatly increased in comparison to pollution standards valid up to 1974.

In the USA, the primary and secondary standards are even more rigorous (JAPCA, 1971). Primary standards are concerned with the protection of humans from harmful pollution effects, while secondary standards are meant to provide adequate protection to vegetation, the most sensitive group affected by SO_2.

Primary Standard

80 μg SO_2/m^3 air (0.03 ppm)	yearly average
365 μg SO_2/m^3 air (0.14 ppm)	maximum 24h average which may be exceeded only once yearly

Secondary Standard

60 μg SO_2/m^3 air (0.02 ppm)	yearly average
260 μg SO_2/m^3 air (0.10 ppm)	maximum 24h average which may be exceeded only once yearly
1300 μg SO_2/m^3 air (0.5 ppm)	maximum 3h average which may be exceeded only once yearly.

In West Germany the following standards for hydrogen fluoride are valid:

$$IW_1 = 2.0\ \mu g\ HF/m^3\ air,$$
$$IW_2 = 4.0\ \mu g\ HF/m^3\ air.$$

At such levels, a considerable risk of injury to many conifers, such as *Picea abies*, *Abies nordmanniana*, *Pinus strobus*, *P. ponderosa*, and *Larix* spp., must be expected. Long-term exposure over a period of years or decades could cause die-out of entire stands. Injury is also possible for many deciduous trees, such as *Fagus silvatica*, *Carpinus betulus*, *Sorbus intermedia*, *Acer palmatum atrop.*, and *Syringa vulgaris*.

Reduction of growth, yield, and quality, especially of stone fruits, such as apricot (*P. armeniaca*), peach (*P. persica*), cherry (*Prunus* spp.) and plum (*P. domestica*), is to be expected. This is also true for currant (*R. rubrum* and *R. nigrum*), gooseberry (*R. grossularia*), strawberry (*F. chiloensis* var. *ananassa*), and grape (*V. vinifera*).

Severe effects on growth and yield of most agricultural plants are not probable, but fluorosis, due to fluoride accumulation in fodder plants is a distinct possibility. Toxic fluoride levels are more likely to occur in meadow and pasture plants than in those cultivted specifically for fodder.

For ornamentals, damage to many tuber and bulb plants must be expected, for example on tulip, gladiolus, daffodil, hyazinth (*Hyazinthus orientalis*), crocus (*Crocus vernus*), scilla (*Scilla sibirica*), tigerlily (*Ferraria pavone*), and begonia (*B. tuberhybrida*). The allowed HF levels also present the risk of injury to many lichen species.

Acording to Weinstein (1971), the following fluoride standards have been valid in the State of Oregon since 1967:

0.65 ppb = 0.50 μg F/m^3 air during the growing period
1 ppb = 0.78 μg F/m^3 air during one month (30 days)
2 ppb = 1.55 μg F/m^3 air during one week (7 days)
3.5 ppb = 2.70 μg F/m^3 air during one day (24 h)
4.5 ppb = 3.50 μg F/m^3 air during one 12h period.

These standards, which have been accepted with modifications by several of the states of the USA, provide more protection for vegetation and animals than the fluoride standards now valid in West Germany.

The allowable limits for HCl (TA Luft, 1974), $IW_1 = 0.10$ mg HCl/m^3, $IW_2 = 0.20$ mg HCl/m^3, are comparable with those for SO_2. This pollutant does not pose as great a hazard however, because HCl usually only occurs locally in concentrations high enough to cause injury to vegetation.

4.2 Plants as Indicators for Pollutants Containing Sulfur, Fluoride, and Chloride

Plants are not only sensitive to the pollutants considered here, but sometimes even show specific responses to and can be used as biological indicators for determination and evaluation of effects of pollutants. Comparative studies are also easier to carry out with plants than with animals or humans, because plants are usually found in large numbers in a specific habitat and are exposed to environmental conditions that are easily determined (Heck, 1966; Brandt and Heck, 1967; Schönbeck et al., 1970; Guderian and Schönbeck, 1971). Differences in the chemical composition and effectiveness of pollutants are expressed in various and, generally, measurable plant reactions.

These methods are biological tests in which the effectiveness of certain substances is determined through responses of living things. Responses of living things, therefore, allow assertions regarding effects, while chemical and physical analysis of the air provide a basis for determining the risk to biological objects. Biological objects also respond only to a specific part of the total pollutant complex.

Indicator plants are useful for the recognition of air pollutants and also for their monitoring and surveillance. For the recognition of phytotoxic air pollutants, one usually relies on observations and studies on wild and cultivated plants which occur in polluted areas (Guderian and van Haut, 1970). An overview of possible methods is presented in Figure 37.

After the possibility that injury comes from biotic or other abiotic agents is eliminated, the detection of air pollutants is carried out through the following types of investigation. The first important evidence is the form of external injury, such as necrosis or changes in growth habit (van Haut and Stratmann, 1970; Jacobson and Hill, 1970). Chemical analysis to determine the accumulation of pollutants in plant material and analysis of the air to determine the presence of pollutant concentrations high enough to cause injury aid in the recognition of potential causes of injury and support the diagnosis. Through exposure of higher and lower indicator plants, even in areas where several pollutants occur at the same time, SO_2, HF, and HCl can also be detected.

Whereas effects on wild and cultivated plants allow the recognition of specific air pollutants, measurement and surveillance of air pollutants is carried out by standardized exposure of indicator plants in the study area. For such studies, the following methods are possible:

1. Exposure of plants in containers or in plots.
2. Exposure in test chambers with filtered and nonfiltered air.
3. Exposure on special stands.
4. Tests of plants in the laboratory.

Particular responses are quantitatively determined for use as values of measurement. These responses include degree of foliar injury, rate of growth, amount of yield, rate of mortality, or accumulation of pollutants in plant organs. In order to quantify and compare such responses, the plant material must not only be genetically uniform, but plants must also be grown under standardized cultivation and exposure conditions.

The standardized grass culture, developed by Scholl (1971b), is particularly suited for surveillance of pollutants in areas with fluoride emissions. For this method, the grass, *L. multiflorum*, is cultivated under standardized conditions, placed in plastic containers that have an automatic watering system, and exposed for two weeks at sites of a special monitoring network (Scholl, 1972).

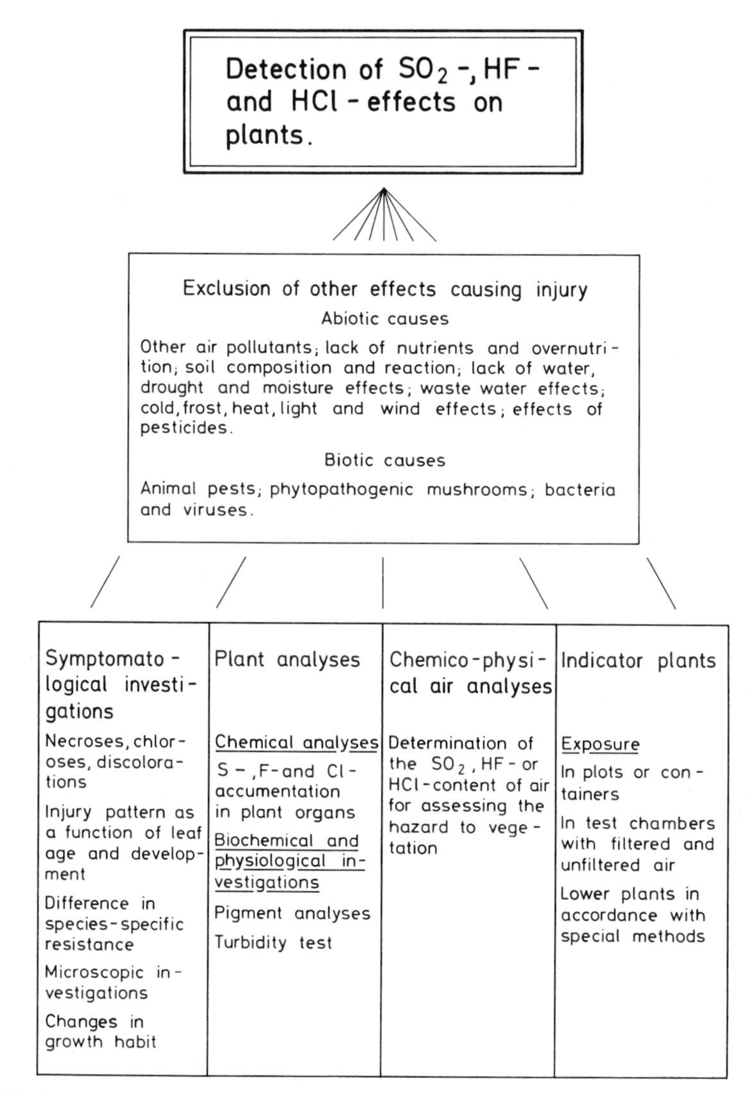

Fig. 37. Methods for the recognition of effects of SO_2, HF, and HCl on vegetation

Grass samples are analyzed for fluoride content, for example with the fluoride-specific electrode (Reusmann and Westphalen, 1969). Fluoride levels in the plant material indicate the fluoride load of the air in the exposure area. The usefulness of the grass culture as an integrated monitoring system has been proven in areas with single fluoride sources as well as in heavily industrialized areas. This method has also been used successfully in areas with pollutants containing sulfur, chloride and heavy metals.

Accumulation of pollutants also occurs in plant products used as food stuffs for humans (Guderian, 1973). Kale *(B. oleracea acephala)*, which was exposed as an indicator plant in the Ruhr area for three months, had a fluoride content of up to 200 ppm in dry matter compared with control values below 20 ppm (van Haut, 1972a). In consideration of normal eating habits, a risk to humans from food stuffs contaminated with fluoride is thought to be slight (Brandt, 1971). Fluoride uptake of humans in the USA, however, has increased in the last two decades from 0.4 mg F/person/day to 1.5 mg (Kintner, 1971). This increase is caused mainly by the high fluoride content of food stuffs from polluted areas or use of fluoridated water in food processing.

The accumulation of fluoride in the plant material of fluoride-resistant indicator plants serves as an indication of the amount of atmospheric fluoride. Plants that are very sensitive to fluoride and show foliar necrosis at low levels of fluoride are also used. Tuberous and bulbous plants are particularly suited to this purpose as they are relatively resistant to SO_2. Spierings (1967) has planted test fields with the sensitive gladiolus variety Snow Princess in various sections of Holland. Tip necrosis and fluoride content of the leaves increased as fluoride concentration in the atmosphere increased. Gladiolus has also been used successfully in the USA and Canada as an indicator plant for fluoride (Adams et al., 1957; Linzon, 1971).

When climatic conditions at the various exposure sites vary extremely or when several pollutants occur together, the use of the exposure methods described above becomes more difficult. These problems can be avoided through comparative studies on the same site with indicator plants in chambers with filtered and nonfiltered air (van Haut, 1972b). The "open-top" chambers developed by Mandl et al. (1973) and Heagle et al. (1973), seem to be well suited for determining pollutant effects on production processes under natural conditions. A large, clear plastic cylinder, open at the top, is placed around the plants to be studied. A constant air stream is pumped into the chamber at ground level. One chamber receives filtered air, the other, nonfiltered air. This system should make it possible to study effects of long-term exposure of ecosystems to low pollutant concentrations.

Along with higher plants, epiphytic organisms, such as lichens, are particularly suitable as indicators of pollutant action (Ferry et al., 1973). Three reasons are given for the decrease in species diversity of lichen populations or for the occurrence of "lichen deserts" in polluted areas (Steiner and Schulze-Horn, 1955; Skye, 1968): the effect of a city climate, characterized by low humidity and reduced dew formation, effects of air pollutants, and the combined effects of both factors. Field studies and fumigation experiments have shown that lichens show measurable responses to air pollutants. Brodo (1966), working with transplanted lichens under natural ecologic conditions, showed that conditions of New York City still had an influence even at a great distance from the city. Kirschbaum et al. (1971) used lichens to determine pollutant stress in Frankfurt/Main. According to Schönbeck (1968), SO_2 emissions from an iron ore smelter were responsible for the disappearance of certain foliose lichens and a general reduction of the total lichen population in the area. Under an average concentration of 0.23 mg SO_2/m^3 air, the test lichen, *Hypogymnia physodes*, died out completely in 29 days. On the site with the lowest concentration, 0.08 mg SO_2/m^3 air, 60% necrosis of the total thallus area occurred after exposure for 68 days. Leblanc (1971) found a good correlation between the degree of injury of the natural lichen flora and the SO_2 concentration in the atmosphere. In fumigation experiments, *H. physodes*

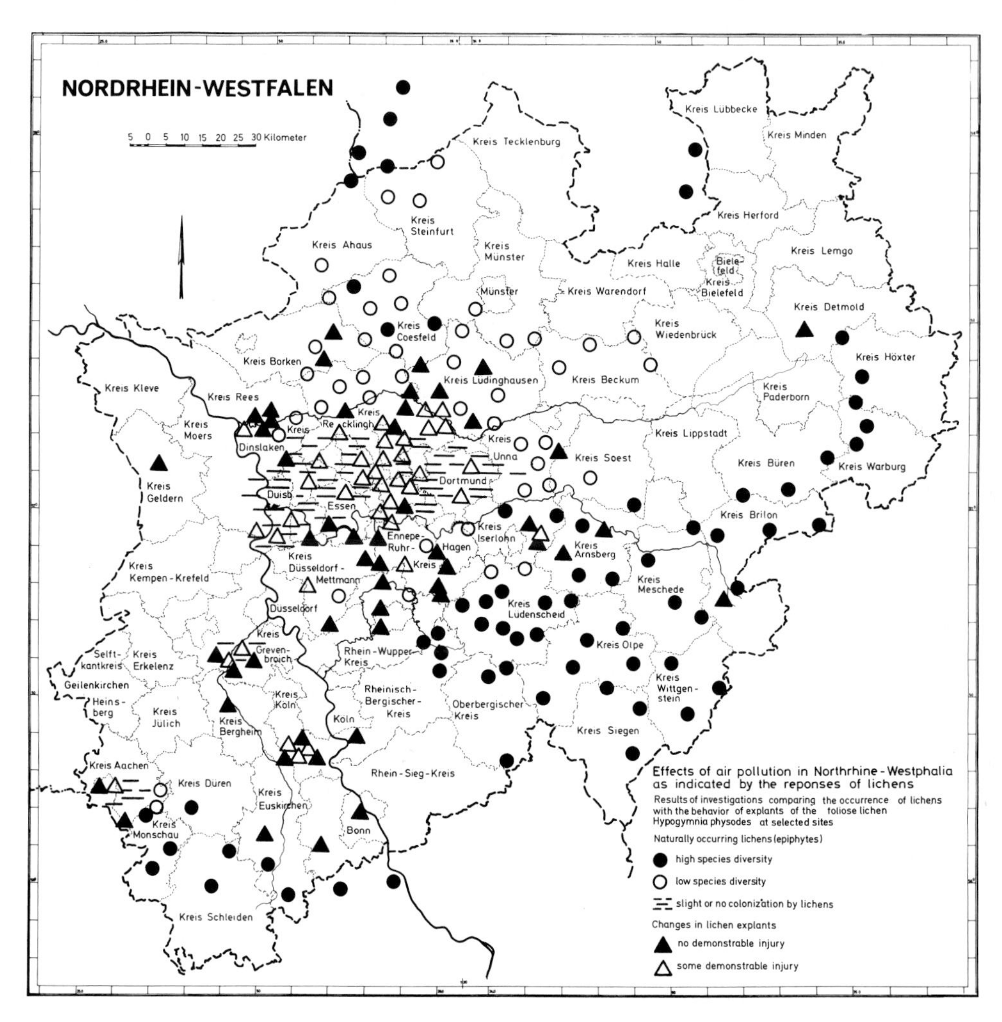

Fig. 38. Influence of air pollution in North Rhine Westphalia based on the reactions of lichens. (After Schönbeck, 1972)

proved to be very sensitive to SO_2, HF, and HCl (Guderian and Schönbeck, 1971).

The sensitivity of lichens to air pollutants led to the study of their possible use as indicators of pollutant effects. The most suitable lichen for this purpose proved to be the foliose lichen *H. physodes* (Schönbeck, 1969), which is widely distributed throughout Europe (Grumann, 1963). Lichens are cut, with the bark, from trees in nonpolluted areas, placed on special stands, and set out in the particular study areas. The death rate of the thallus is determined by photographs taken at definite time intervals. In this way pollutant conditions can be monitored and the extent of the polluted area can be defined, as experiments near single sources and in large industrial areas have shown. For example, the death rate of *Hypogymnia* near an aluminum smelter depended on the amount of atmospheric fluoride (Guderian and Schönbeck, 1971). Exposure of this lichen in and near the Ruhr area showed the usefulness of this method for monitoring pollutants over large areas (Fig. 38). With increasing distance from the center of the Ruhr area, death rate of the test lichen decreased and species diversity of the total lichen population increased (Schönbeck, 1972).

The responses of lichens do not yet allow a direct determination of the risks to higher plants. To determine the correlation between reactions of lichens and potential risks to higher plants, comparative experiments with lichens and higher plants were carried out in fumigation chambers with SO_2, HF, and HCl (Schönbeck and Guderian, 1976). The following trends could be recognized: under concentrations of SO_2, HF, or HCl that cause acute injury to higher plants, *Hypogymnia* died within weeks or months. Compared with higher plants, the lichen responded quite slowly, as can be seen from a fumigation experiment with 1.9 mg SO_2/m^3 air (Fig. 14). Necrosis of higher plants occurred before any external reaction of the lichen thallus could be detected. Short-term exposure to high pollutant concentrations appear to be less dangerous for *Hypogymnia* than for certain higher plants. Long-term exposure to low concentrations, however, causes more lasting injury to the lichen than to higher plants as is particularly clearly shown by the experiment presented in Figure 39.

Further long-term fumigation experiments, with very low concentrations, as well as field studies (Schönbeck and Guderian, 1976), have shown that *Hypogymnia* dies under concentrations which cause injury only to the most sensitive higher plants. *H. physodes*, therefore, is a good indicator for low pollutant concentrations in the atmosphere. From the death rate of foliose lichens exposed in a standardized manner, conclusions can be made regarding potential effects to higher plants. A definite relationship between the degree of injury of the natural lichen population and of stands of *Pinus silvestris* could be made (Schönbeck, 1972; Knabe, 1972a).

Sensitivity of lichens to air pollutants depends on several factors. Injury, which is longer lasting than that to higher plants, results partly from the poor regenerative ability of lichens. Higher plants, especially annuals, can compensate for the loss of assimilatory tissue through formation of new organs which are generally less sensitive (Fig. 14). Injury to lichens under long-term pollutant stress, however, continues until death of the lichen thallus.

Moisture is also a determining factor in susceptibility of lichens. Fumigation experiments and field studies on moist and dry thalli of *H. physodes* indicate that long-term moisture deficiency increases sensitivity of the lichen (Schönbeck and Guderian, 1976), probably through reduced vitality. Pollutant stress, in connection with dryness, is a reason for the disappearance of this lichen from cities. On the other hand, higher pollutant concentrations that cause injury after short exposure reduce pollutant uptake and effects, with decreasing water potential of the lichen thalli, as has also been shown in studies by Türk et al. (1974).

Pollutant accumulation in lichens is comparable with that in higher plants (Fig. 40). Under various SO_2 concentrations, sulfur accumulation rates were found in lichens which were similar to those of strawberry and white pine. Comparison of the sulfur uptake with concentration *(c)* and exposure time *(t)* under

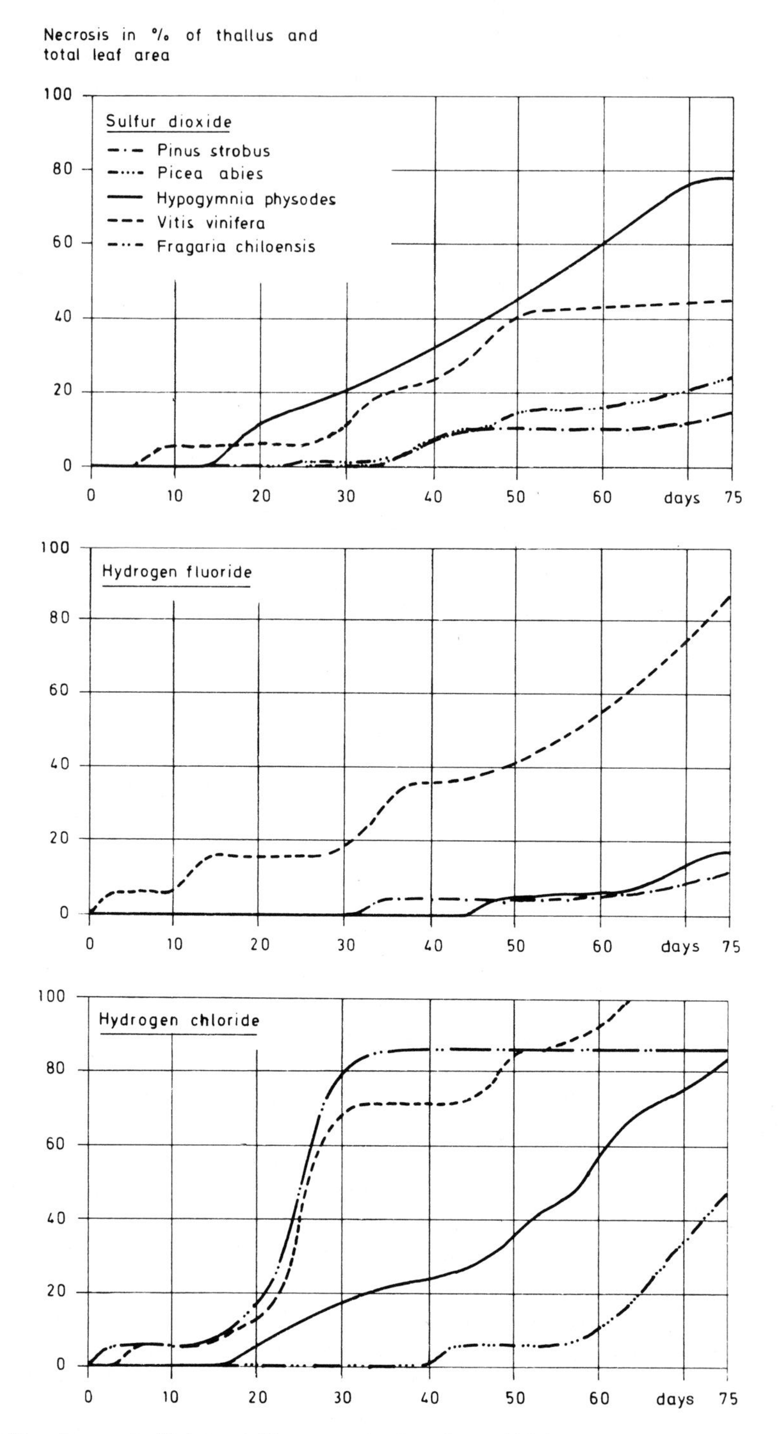

Fig. 39. Development of injury on *Hypogymnia physodes* and higher plants exposed to 0.42 mg SO_2, 2.1 µg HF, and 0.25 mg HCl/m^3 air

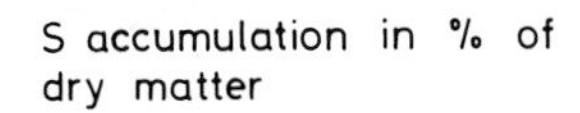

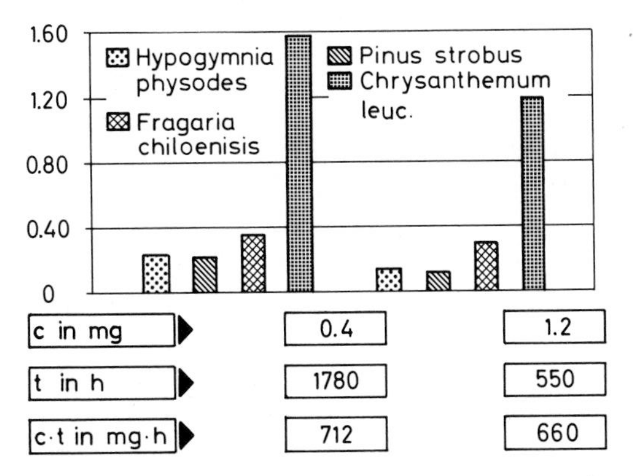

Fig. 40. Sulfur accumulation in *Hypogymnia physodes* and higher plants exposed to SO_2

similar products of *c* and *t* showed that pollutant uptake of lichens was more dependent on exposure time than on concentration.

There were also no major differences in pollutant accumulation in thalli of *Hypogymnia* or leaves of higher plants under HF or HCl concentrations. As described in Section 2.5.3, however, the same amounts of pollutants in plant organs has a greater effect when metabolic activity is low. It follows, then, that at the same pollutant uptake rates, there is more risk of injury to lichens than to higher plants, in part because lichens lack the ability to dilute the absorbed pollutants through formation of large amounts of new plant material with low natural levels of the particular element in question.

4.3 Measures for the Reduction of Pollutant Action in Plant Stands

In the preceeding sections, it has been shown that an increase in pollutant stress which presents a hazard to vegetation over large areas is to be expected. Air pollution studies, therefore, should be systematically included in investigations of population dynamics and synecology. By means of such studies can be determined which changes in interspecific interactions and in the relationships between living things and their environment in single terrestrial and aquatic ecosystems are caused by the action of atmospheric pollutants. The "lichen deserts" of large cities and industrial areas, as well as the reduction of species diversity of lichen populations in less polluted areas, provide evidence of the very sensitive reactions of some groups of organisms to external influences. Systematic monitoring of the flora and fauna over large areas becomes necessary as human activity leads to changes in bio–geochemical cycles which influence the composition of the atmosphere world-wide (Heggestad and Heck, 1971).

In agriculture, gardening, and forestry, air pollution must be considered as a site factor that affects quantity and quality of the particular crop. Injury to vegetation from air pollution cannot always be avoided by control measures at the growing site, but it can at least be moderated. In the development and use of such measures it should not be forgotten that not only are agricultural losses reduced, but a diverse and strong vegetation also contributes to the improvement of air quality through filter effects and dilution of pollutants through increased turbulence. The following possibilities for the reduction of pollutant effects through cultivation practices are derived from species-specific resistance, and from the modifying influence of internal and external growth factors on plant reactions.

4.3.1 Cultivation of Resistant Species and Varieties

Because of the large differences in resistance, cultivation of resistant species represents the greatest possibility for reducing injury due to air pollutants. Since the degree of resistance is strongly dependent on pollutant type, exact knowledge of the pollutant type and concentrations that occur is necessary for success of this method. Especially where perennials are grown in polluted areas, systematic investigations of soil and climate, as well as of air pollution conditions (Guderian, 1969) should be carried out. Methods for these studies are presented in Sections 1.2, 3.2, and 4.2. From results of such studies and consideration of species-specific resistance, possible risks of injury to plants from air pollutants can be determined.

The following summary (Table 26) shows which risks exist for the various branches of agricultural land-use from the mainly chronic effects of SO_2, HF, and HCl. Groups are based on reduction of plant productivity or function which has been reported in the literature or come from personal investigations and are necessarily quite broad to enable better comparison and provide a better survey. Therefore, only a framework, which can be expanded by more detailed investigations, is presented. The classification is based solely on cultivation aspects. Economic aspects are not considered, because whether or not cultivation of particular species under pollutant stress is economically sound depends on the economic management of each single case.

As shown in Table 26 there are different risks from chronic effects of air pollutants for the various land-use practices. All three pollutants pose a great risk for fruit trees and economically important conifers. Under conditions that are not too severe, sensitive conifers are replaced with deciduous trees, a practice that leads, however, to a large loss in yield (Wentzel, 1963, 1968). Higher pollutant stress makes a complete change of culture necessary in forestry and fruit tree cultivation.

Agronomic and ornamental plants are, for the most part, in less danger from chronic pollutant effects. There are gradual differences between the effects of the three components, however. For example, fluoride concentrations under 1 μg HF/m^3 air caused fluoride accumulation in fodder and meadow plants that was dangerous to animals, as described in Section 4.1.1.2.

Through proper crop selection, a particular type of culture can be adapted to air pollution conditions of the growing area. Relatively resistant plants for fodder and field vegetable cultures include cabbage and kale species. Resistant ornamentals include members of the Aralaceae, Ericaceae, and Compositae. Cultivation of

Table 26. Degree of risk to single branches of agricultural land-use from chronic effects of SO_2, HF, and HCl[a]

Plant species	SO_2	HF	HCl
Agriculture			
Grain, including corn	◔	◔	◐
Legumes, clover, alfalfa	◐	◐	◐
Cruciferae, sunflower	○	○	◔
Potato	○	○	◔
Beta and Brassica beets	○	◐	○
Fodder plants[b]			
Grain and grass species	◔	●	◐
Fodder legumes including cloverlike fodder plants	◐	●	◐
Cruciferae and sunflower	○	●	◔
Cabbage and kale species	○	●	○
Field vegetables			
Cruciferae	○	◔	○
Papilionaceae	◐	◐	◐
Umbelliferae	○	○	○
Chenopodiaceae	◕	◐	◐
Cucurbitaceae	○	○	
Compositae	◐	◔	
Solanaceae	○	○	◔
Liliaceae	○	●	◔
Meadows and pastures			
Mainly grass species	◔	●	◐
With large percentage of cloverlike plants	◐	●	◐
Fruit			
Seed fruits	◕	◕	●
Stone fruits	◔	●	◐
Berries, walnut	●	●	◕
Hazelnut	◐	●	●
Grape	◐	●	●
Strawberry	○	●	●
Forests and woods			
Fir, Spruce, White pine, Douglas fir	●	●	●
Austrian pine, Thuja, Yew, Chamaecyparis, Juniper, Deciduous trees	◐	◐	◐
Ornamentals			
Liliaceae, Iridaceae, Amaryllidaceae	○	●	◔
Ranunculaceae, Rosaceae	◔	◔	◔
Papilionaceae	◐	◐	◐
Geraniaceae	○	○	
Araliaceae	○	○	
Caryophyllaceae	◔	◐	
Ericaceae	○	○	○
Compositae	○	○	○

[a] ○ Very slight, ◔ slight, ◐ medium, ◕ severe, ● very severe.
[b] For green fodder, hay, and silage.

stone fruits in areas with fluoride pollution and seed fruits in areas with SO_2 pollution is not recommended. Berry fruits are sensitive to all three pollutants.

The effects of external growth factors on the degree of sensitivity is discussed in Section 2.5. In polluted areas, habitat conditions must be considered in the proper selection of crops for cultivation, remembering, for example, that pollutants can reduce the frost-hardiness of plants.

The number of species that can be cultivated is reduced greatly as pollutant stress increases (Guderian and Stratmann, 1968). Reduced possibilities for crop rotation and adaptation to soil and climate present great economic problems. Combined with difficulties in the optimal marketing of agricultural products, these problems sometimes lead to conditions that threaten the financial existence of the grower.

4.3.2 Fertilizing and Cultivation

Fertilizing. As a measure for reducing action of air pollutants, fertilizing improves the negative effects of the soil and has a direct effect on the progression of injury to above-ground plant organs.

Extreme acidification of soils can occur under the influence of SO_2 (Guderian and Stratmann, 1962). Because of the dependence of vegetation on the soil reactivity (Boeker, 1964), alterations in species diversity of meadow land and natural ecosystems can occur. For agricultural areas it is important that enough calcium be present for maintenance of the optimal pH range, ion-exchange reactions, colloid saturation, and friability of the soil. Calcium, as a nutrient, also increases resistance to SO_2 and HF, as shown in Section 2.5.2.2.

This is also true for nitrogen. For certain plants, optimal nitrogen conditions reduced uptake of SO_2 and HF. Directly following heavy application of nitrogen, however, sensitivity can increase and gradual fertilizing is recommended.

Potassium also increases resistance. Especially under SO_2, an optimal supply of this nutrient reduces the extent of injury, as the uptake rate of SO_2—as opposed to that of HF—is not raised.

An optimal supply of phosphorous is particularly important in areas with fluoride pollution. In fertilizing, it must be considered, however, that use of certain complex fertilizers containing boron, which are produced through breakdown of a phosphate-potassium-boron mixture with strong acids, can lead to an increase in fluoride accumulation in leaves (Bolay et al., 1971 b).

In general, fertilizing should be carried out so as to provide the least possible amounts of those elements that occur as pollutants. Furthermore, fertilizers should be given in chemical forms that reduce uptake of these particular elements. For example, since NO_3^- ions supress uptake of Cl^- ions (Geissler, 1953), it should follow that, under HCl, such a nitrate fertilizer will have a greater effect than one with ammonium nitrogen.

Cultivation and Stand Structure. Solitary trees and shrubs, and those on the borders of stands, are more subject to pollutant effects than those within a stand. An early closing of the stand, therefore, provides a certain amount of protection. Recent investigations (Ulrich, 1972) have shown that this is not only due to changes in flow conditions, but also to filter effects of leaves, branches, and stems.

Penetration of pollutants into the stand can be hindered by reducing the distance between individuals and by less extensive thinning. The phenotypically resistant individuals in forest stands injured by pollutants should be protected regardless of their sociologic position in the community (Wentzel, 1963).

Through use of various pruning and thinning techniques (Hilkenbäumer, 1964), a compact crown of fruit trees should be promoted in polluted areas, even when other disadvantages result. As a result of reduced branch growth under polluted conditions, early senescence commonly occurs at the beginning of the yield stage (Guderian, 1969). Pruning is of particular importance, therefore, for rejuvenating vegetative and generative growth.

Differences in plant resistance also make it possible to use planting of resistant plants for protection of more sensitive cultures. Depending on topography and pollution conditions, protective rows of resistant plants should be planted before the sensitive crop. Especially with low-growing fruits and berries, protective rows of less sensitive fruits are possible, for example pears and prunes in areas under the influence of SO_2. The alternate cultivation of low and tall growing cultures, as recommended for wind protection can also be advantageous.

References

Adams, D. F.: Recognition of the effects of fluorides on vegetation. JAPCA **13**, 360–362 (1963)

Adams, D. F., Hendrix, J. W., Applegate, H. G.: Relationship among exposure periods, foliar burn and fluorine content of plants exposed to hydrogen fluoride. J. Agric. Food Chem. **5**, 108–116 (1957)

Adams, D. F., Shaw, C. G., Gnagy, R. M., Koppe, R. K., Mayhew, D. J., Yerkes, W. D.: Relationship of atmospheric fluoride levels and injury indexes on gladiolus and ponderosa pine. J. Agric. Food Chem. **4**, 64–66 (1956)

Alvik, G.: Über Assimilation und Atmung einiger Holzgewächse im westnorwegischen Winter. Medd. **22** Vestlandets Forst. Forsökssta. Bergen, 1939

Applegate, H. G., Adams, D. F.: Invisible injury of bush beans by atmospheric and aqueous fluorides. Intern. J. Air Water Pollution **3**, 231–241 (1960)

Applegate, H. G., Durrant, L. C.: Synergistic action of ozone-sulfur dioxide on peanuts. Environ. Sci. Technol. **3**, 759–760 (1969)

Arndt, U.: Konzentrationsänderungen bei Blattfarbstoffen unter dem Einfluß von Luftverunreinigungen. Ein Diskussionsbeitrag zur Pigmentanalyse. Environ. Pollution **2**, 37–48 (1971)

Arndt, U., Stramplat, W., Krautscheid, S.: Aufbau und Funktion eines Klimakammersystems für Begasungsexperimente. Essen: Girardet-Verlag, Schriftenr. Landesanst. Immissions-Bodennutzungssch. d. Landes Nordrhein-Westfalen **29**, 49–57 (1973)

Arnold, A.: Die Bedeutung der Chlorionen für die Pflanze. Jena: Bot. Stud. H. **2**, 1955

Arnon, D. J.: Conversion of light into chemical energy in photosynthesis. Nature **184**, 10–21 (1959)

Auersch, O.: Zur Beeinflussung der Obstgehölze durch Immissionen. Erwerbsobstbau **9**, 153–156 (1967)

Bassham, J. A., Benson, A. A., Kay, L. D., Harris, A. Z., Wilson, A. T., Calvin, M.: The path of carbon in photosynthesis. XXI. The cyclic regeneration of carbon dioxide acceptor. In: Calvin, M., Bassham, J. B. (eds.) The Photosynthesis of Carbon Compounds. New York: W. A. Benjamin Inc., 1962

Baumeister, W., Burghardt, H.: Die Bedeutung der Elemente Zink und Fluor für das Pflanzenwachstum. Forschungsber. Wirtsch. Verkehrsministeriums Nordrhein-Westfalen Nr. **388** (1957)

Bell, J. N. B., Clough, W. S.: Depression of yield in ryegrass exposed to sulphur dioxide. Nature **241**, 47–49 (1973)

Benedict, H. M., Ross, J. M., Wade, R. W.: The disposition of atmospheric fluorides by vegetation. Intern. J. Air Water Pollution **8**, 279–289 (1964)

Benedict, H. M., Ross, J. M., Wade, R. W.: Some responses of vegetation to atmospheric fluorides. JAPCA **15**, 253–255 (1965)

Bernatzky, A.: Schutzpflanzungen zur Luftreinigung und Besserung der Umweltbedingungen. Baum-Ztg. **2** (3), 37–42 (1968)

BImSchG: Gesetz zum Schutz vor schädlichen Umwelteinwirkungen durch Luftverunreinigungen, Geräusche, Erschütterungen und ähnliche Vorgänge (Bundes-Immissionsschutz-Gesetz — BImSchG vom 15. März 1974). Bonn: BGBl. I S 721–743 (1974)

Bleasdale, J. K. A.: Atmospheric pollution and plant growth. Nature **169**, 376–377 (1952)

BMI: Umweltschutz. Das Umweltprogramm der Bundesregierung. Stuttgart, Berlin, Mainz, Köln: W. Kohlhammer 1972
Bochow, H.: Bodenmüdigkeit. In: Phytopathologie und Pflanzenschutz I, Berlin: Akademie-Verlag, 1965, pp. 292–300
Boeker, P.: Die Verbreitung der wichtigsten Grünlandpflanzen Nordrhein-Westfalens in Abhängigkeit vom pH-Wert. Hiltrup: Landwirtschaftsverlag, Forsch. Berat. Reihe B, **10**, 211–230 (1964)
Bolay, A., Bovay, E.: Observations sur la sensibilté aux gaz fluorés de quelques espèces vegetales du valais. Phytopathol. Z. **53**, 289–298 (1965)
Bolay, A., Bovay, E., Neury, G., Quinche, J. P., Zuber, R.: Dégats causés aux abricots et à d'autres fruits par les composés fluorés. Tirage Revue Suisse viticult. arboricult. **3**, 82–92 (1971 a)
Bolay, A., Bovay, E., Quinche, J. P., Zuber, R.: Teneurs en fluor et en bore des feuilles et des fruits d'arbres fruitiers et de vignes, fumes avec certain engrais composés, boriqués fluorés. Tirage Revue Suisse viticult. arboricult. **3**, 54–61 (1971 b)
Boresch, K.: Weitere Untersuchung der durch Chloride hervorgerufenen Blattrandkrankheit der Johannisbeere. Bodenk. Pflanzenernähr. **14**, 230–247 (1939)
Borgsdorf, W.: Beiträge zur Fluorschadendiagnostik I. Fluorschaden — Weiserpflanzen in der Wildflora. Phytopathol. Z. **38**, 309–315 (1960)
Bösener, R.: Zum Vorkommen rindenbrütender Schadinsekten in rauchgeschädigten Kiefern- und Fichtenbeständen. Arch. Forstwes. **18**, 1021–1026 (1969)
Brandt, C. S.: Effect of air pollution on plants. In: Air Pollution. New York: Academic Press, 1962, Vol. **1**, pp. 255–281
Brandt, C. S.: Ambient air quality criteria for hydrofluorine and fluorides. Düsseldorf: VDI-Verlag, VDI-Berichte **164**, 23–27 (1971)
Brandt, C. S., Heck, W. W.: Effects of air pollution on vegetation. In: Air Pollution. New York: Academic Press, 1967, Vol. **1**, pp. 401–443
Brandt, C. J., Rhoades, R. W.: Effects of limestone dust accumulation on composition of a forest community. Environ. Pollution **3**, 217–225 (1972)
Bredemann, G.: Biochemie und Physiologie des Fluors. Berlin: Akademie-Verlag, 1956
Brennan, E., Leone, I. A., Daines, R. H.: Fluorine toxicity in tomato as modified by alterations in the nitrogen, calcium and phosphorus nutrition of the plant. Plant Physiol. **25**, 736–747 (1950)
Brennan, E., Leone, I. A., Daines, R. H.: Chlorine as a phytotoxic air pollutant. Intern. J. Air Water Pollution **9**, 791–797 (1965)
Brennan, E., Halisky, P. M.: Response of turfgrass cultivars to ozone and sulfur dioxide in the atmosphere. Phytopathologia **60**, 1544–1546 (1970)
Brewer, R. F., Sutherland, F. H., Guillemet, F. B., Creveling, R. K.: Some effects of hydrogen fluoride gas on bearing navel orange trees. Proc. Am. Soc. Hort. Sci. **76**, 208–214 (1960 a)
Brewer, R. F., Sutherland, F. H., Guillemet, F. B., Creveling, R. K.: Sorption of fluorine by citrus foliage from equivalent solutions of HF, NaF, NH_4F, and H_2SiF_6. Proc. Am. Soc. Hort. Sci. **76**, 215–219 (1960 b)
Brisley, H. R., Jones, W. W.: Sulfur dioxide fumigation of wheat with special reference to its effects on yield. Plant Physiol. **25**, 666–681 (1950)
Brocke, W., Schade, H.: Die Luftverunreinigung durch Abgase aus der Verbrennung von Brennstoffen in stationären Anlagen der Bundesrepublik Deutschland. Staub Reinh. Luft **31**, 473–478 (1971)
Brodo, J. M.: Lichen growth and cities: A study on Long Island, New York. Bryology **69**, 427 (1966)
Buck, M.: Die Bedeutung des Immissionsschutzes für die Vegetation. Staub Reinh. Luft **30**, 26–31 (1970)
Buck, M., Reusmann, G.: A new semi-automatic method for fluoride determination in plant and air samples. Fluoride Quart. Rep. **4**, No. 1, 5–15 (1971)
Buck, M., Stratmann, H.: Ein Verfahren zur Bestimmung sehr geringer Konzentrationen von Fluor-Ionen in der Atmosphäre. Brennstoff-Chem. **46**, 231–235 (1965)
Burghardt, H.: Über die Bedeutung des Chlors für die Pflanzenernährung unter besonderer Berücksichtigung des Chlorid/Sulfat-Problems. Angew. Botan. **36**, 203–257 (1962)

Cantwell, A. M.: Effect of temperature on response of plants to ozone as conducted in a specially designed plant fumigation chamber. Plant Disease Reptr. **52**, 57–69 (1968)

Cernusca, A.: Ökophysik: Neue Wege zur quantitativen Ökologie. Umschau **71**, 663–668 (1971)

Chang, W., Thompson, C. R.: Subcellular distribution of fluoride in navel orange leaves. Intern. J. Air Water Pollution **9**, 685–691 (1965)

Compton, O. C., Remmert, C. F.: Effect of air-borne fluorine on injury and fluorine content of gladiolus leaves. Proc. Am. Soc. Hort. Sci. **75**, 663–675 (1960)

Dässler, H. G., Grumbach, H.: Abgasschäden an Obst in der Umgebung eines Fluorwerkes. Arch. Pflanzenschutz **3**, 59–69 (1967)

Dean, R. S., Swain, R. E.: Report submitted to the Trail Smelter Arbitral Tribunal. Bull. Mines **453**, US Printing Office, 1939

Desai, M. C.: Effects of certain nutrient deficiencies on stomatal behavior. Plant Physiol. **12**, 253–283 (1937)

Deutsche Forschungsgemeinschaft: Mitteilung I v. 1. Juni 1964, Deut. Forschungsgem. Bad Godesberg, 18 S., 1964

Dochinger, L. S., Seliscar, C. E.: Results from grafting chlorotic dwarf and healthy eastern white pine. Phytopathology **55**, 404–407 (1965a)

Dochinger, L. S., Seliscar, C. E., Bender, F. W.: Etiology of chlorotic dwarf of eastern white pine. Phytopathology **55**, 1055 (abstracts), 1965b

Domes, W.: Unterschiedliche CO_2-Abhängigkeit des Gasaustausches beider Blattseiten von *Zea mays*. Planta **98**, 186–189 (1971)

Donaubauer, E.: Durch Industrieabgase bedingte Sekundärschäden am Wald. In: Forstliche Rauchschäden in Österreich. Mitt. Forstl. Bundesversuchsanst. Mariabrunn. Wien: Österr. Agrarverlag, 1966, Vol. LXXIII, pp. 101–110

Dörries, W.: Über die Brauchbarkeit der spektroskopischen Phäophytinprobe in der Rauchschaden-Diagnostik. Z. Pflanzenkrankh. Pflanzenschutz **42**, 257–273 (1932)

Dreisinger, B. R.: Monitoring atmospheric sulphur and correlating its effects on crops and forests in the Sudbury area. Impact of Air Pollution on Vegetation Conference Park Plaza Hotel, Toronto, Ontario 7.–9. April, 1970

Dreyhaupt, F. J.: Luftreinhaltung als Faktor der Stadt- und Regionalplanung. Köln: Verlag TÜV Rheinland und Heymanns, 1971, Schriftenr. Umweltschutz **1**

Dugger, W. M., Taylor, O. C., Cardiff, E., Thompson, C. R.: Relationship between carbohydrate content and susceptibility of pinto bean plants to ozone damage. Proc. Am. Hort. Sci. **81**, 304–315 (1962)

Dugger, W. M., Taylor, O. C., Klein, W. H., Shropshire, W.: Action spectrum of peroxyacetyl nitrate damage to bean plants. Nature **198**, 75–76 (1963)

Dugger, W. M., Ting, J. P.: The effect of peroxyacetyl nitrate: Photoreductive reactions and susceptibility of bean plants to PAN. Phytopathology **58**, 102–107 (1968)

Egle, K.: Methoden der Photosynthesemessungen. — Landpflanzen. In: Handbuch d. Pflanzenphysiologie. Berlin: Springer, 1960, Vol. I, pp. 115–163

Egnér, H.: Die Bedeutung der Nährstoffzufuhr durch Luft und Niederschläge für die Bodenfruchtbarkeit. Sonderh. Z. Landwirtsch. Forsch. **7**, 90–94 (1956)

Ellenberg, H.: Ziele und Stand der Ökosystemforschung. In: Ökosystemforschung. Berlin, Heidelberg, New York: Springer, 1973, pp. 1–31

Enderlein, H., Kästner, W.: Welchen Einfluß hat der Mangel eines Nährstoffes auf die SO_2-Resistenz 1 jähriger Kiefern. Arch. Forstwes. **16**, 431–435 (1967)

Engle, R. L., Gabelman, W. H.: Response of *Allium cepa* L. to ozone. Proc. 17. Intern. Hort. Congr. Univ. Maryland 1966/1, No. 472, 1966

Eriksson, E.: Composition of atmospheric precipitation. Tellus **4**, 215–232 and 280–303 (1952)

Ewert, E., Dässler, H. G.: Schädigung von Pflanzen durch Chlorwasserstoff. Umschau **25**, 839–841 (1969)

Faller, N.: Der Schwefeldioxidgehalt der Luft als Komponente der Schwefelversorgung der Pflanze. Diss., Gießen, 1968

Faller, N.: Schwefeldioxid aus der Luft als entscheidende Nährstoffquelle für die Pflanze. Landwirtsch. Forsch. **18** (25), 48–54 (1970)

Faller, N., Höfner, W.: Schwefelassimilation durch zerkleinertes Pflanzenmaterial nach SO_2-Begasung. Z. Pflanzenernähr. Düng. Bodenk. **121**, 111–116 (1968)
Ferry, B. W., Baddely, M. S., Hawksworth, D. L. (eds.): Air Pollution and Lichens. London: Athlone Press, 1973
Fischer, K., Kramer, D., Ziegler, H.: Elektronenmikroskopische Untersuchungen SO_2-begaster Blätter von *Vicia faba*. I. Beobachtungen an Chloroplasten mit akuter Schädigung. Protoplasma **76**, 83–96 (1973)
Franke, W.: Über die Beziehungen der Ektodesmen zur Stoffaufnahme durch Blätter. III. Nachweis der Beteiligung der Ektodesmen an der Stoffaufnahme durch Blätter mittels radioaktiver Stoffe. Planta **61**, 1–16 (1964)
Franke, W.: Mechanism of foliar penetration of solutions. Ann. Rev. Plant Physiol. **18**, 281–298 (1967)
Freeland, R. O.: Apparent photosynthesis in some conifers during winter. Plant Physiol. **19**, 179–185 (1944)
Freitag, V., Oelschläger, W., Loeffler, K.: Fluoride content and microradiographic findings in skeletal fluorosis. Fluoride **3**, 167 (1970)
Freney, J. R., Barrow, N. J., Spencer, K.: A review of certain aspects of sulphur as a soil constituent and plant nutrient. Plant Soil **17**, 295–308 (1962)
Fuchs, W. H., Rosenstiel, K.: Ertragssicherheit. In: Handbuch der Pflanzenzüchtung. Berlin, Hamburg: P. Parey-Verlag, 1958, Vol. I, pp. 365–442
Garber, K.: Luftverunreinigung und ihre Wirkungen. Berlin: Gebr. Borntraeger Verlag, 1967
Garber, K., Guderian, R., Stratmann, H.: Untersuchungen über die Aufnahme von Fluor aus dem Boden durch Pflanzen. Qualitas Plant. Mater Vegetabiles XIV (3), 223–236 (1967)
Gäumann, E.: Pflanzliche Infektionslehre. Basel: Birkhäuser-Verlag, 1951
Geissler, Th.: Über die Wirkung chlorid- und sulfathaltiger Düngemittel auf den Ertrag einiger Gemüsearten unter verschiedenen Umweltverhältnissen. Arch. Gartenbau **1**, 233–243 (1953)
Grafe, P.: Untersuchungen zur Aufnahme von Fluorwasserstoff durch Blätter höherer Pflanzen. Diplom-Arbeit, Botanisches Inst. Universität Gießen, 1973
Grossmann, F.: Einfluß der Ernährung der Pflanzen auf den Befall durch Krankheitserreger und Schädlinge. Landwirtsch. Forsch. **18** (25), 1970
Grosso, J. J., Menser, H. A., Hodges, G. H., McKinney, H. H.: Effects of air pollutants on *Nicotiana* cultivars and species used for virus studies. Phytopathology **61**, 945–950 (1971)
Grumann, V.: Catalogus Lichenum Germaniae. Stuttgart: G. Fischer Verlag, 1963
Guderian, R.: Zur Methodik der Ermittlung von SO_2-Toleranzgrenzen für land- und forstwirtschaftliche Kulturen im Freilandversuch Biersdorf (Sieg). Staub **20**, 334–337 (1960)
Guderian, R.: Luftverunreinigungen und Pflanzenschutz. Z. Pflanzenkrankh. Pflanzenschutz **73**, 241–265 (1966a)
Guderian, R.: Reaktionen von Pflanzengemeinschaften des Feldfutterbaues auf Schwefeldioxideinwirkungen. Essen: Girardet-Verlag, Schriftenr. Landesanst. Immissions- Bodennutzungssch. d. Landes Nordrhein-Westfalen **4**, 80–100 (1966b)
Guderian, R.: Obstbau in Gebieten mit Schwefeldioxid-Immissionen. Erwerbsobstbau **11**, 110–113 (1969)
Guderian, R.: Untersuchungen über quantitative Beziehungen zwischen dem Schwefelgehalt von Pflanzen und dem Schwefeldioxidgehalt der Luft. Z. Pflanzenkrankh. Pflanzenschutz **77**, I. Teil 200–220, 1970, II. Teil 289–308, 1970, III. Teil 387–399 (1970)
Guderian, R.: Einfluß der Nährstoffversorgung auf die Aufnahme von Schwefeldioxid aus der Luft und auf die Pflanzenanfälligkeit. Essen: Girardet-Verlag, Schriftenr. Landesanst. Immissions-Bodennutzungssch. d. Landes Nordrhein-Westfalen **23**, 51–57 (1971)
Guderian, R., van Haut, H.: Nachweis von Schwefeldioxid-Wirkungen an Pflanzen. Staub Reinh. Luft **30**, 17–26 (1970)
Guderian, R., van Haut, H., Stratmann, H.: Probleme der Erfassung und Beurteilung von Wirkungen gasförmiger Luftverunreinigungen auf die Vegetation. Z. Pflanzenkrankh. Pflanzenschutz **67**, 257–264 (1960)
Guderian, R., van Haut, H., Stratmann, H.: Experimentelle Untersuchungen über pflanzenschädigende Fluorwasserstoff-Konzentrationen. Köln und Opladen. Westdeutscher Verlag. Forsch. Ber. d. Landes Nordrhein-Westfalen, Nr. **2017**, 1969

Guderian, R., Schönbeck, H.: Recent results for recognition and monitoring of air pollutants with the aid of plants. New York and London. Academic Press. Proc. 2nd Intern. Clean Air Congr. 266–273 (1971)

Guderian, R., Stratmann, H.: Freilandversuche zur Ermittlung von Schwefeldioxidwirkungen auf die Vegetation. I. Teil: Übersicht zur Versuchsmethodik und Versuchsauswertung. Köln und Opladen. Westdeutscher Verlag. Forsch. Ber. d. Landes Nordrhein-Westfalen, Nr. **1118**, 1962

Guderian, R., Stratmann, H.: Freilandversuche zur Ermittlung von Schwefeldioxidwirkungen auf die Vegetation. III. Teil: Grenzwerte schädlicher SO_2-Immissionen für Obst- und Forstkulturen sowie für landwirtschaftliche und gärtnerische Pflanzenarten. Köln und Opladen. Westdeutscher Verlag, Forsch. Ber. d. Landes Nordrhein-Westfalen Nr. **1920**, 1968

Guderian, R., Thiel, K.: Versuchsanlage zur Ermittlung immissionsbedingter Kombinationswirkungen an Pflanzen. Essen: Girardet-Verlag, Schriftenr. Landesanst. Immissions- Bodennutzungssch. d. Landes Nordrhein-Westfalen **29**, 61–64 (1973)

Haagen-Smit, A. J., Darley, E. F., Zaitling, M., Hull, H., Noble, W.: Investigation on injury to plants from air pollution in Los Angeles area. Plant Physiol. **27**, 18–34 (1952)

Halbwachs, G.: Untersuchungen über gerichtete aktive Strömungen und Stofftransporte im Blatt. Flora **153**, 333–357 (1963)

Halbwachs, G., Kisser, J.: Durch Rauchimmissionen bedingter Zwergwuchs bei Fichte und Birke. Centralbl. ges. Forstwesen **84**, 156–173 (1967)

Harrison, B. F., Thomas, M. D., Hill, G. R.: Radioautographs showing the distribution of sulphur in wheat. Plant Physiol. **19**, 245–257 (1944)

Hartkamp, H.: Kapillardosierer für die Herstellung primärer Standards zu Eich- und Prüfzwecken durch Absolutdosierung kleinster Gas- und Flüssigkeitsmengen. Essen: Girardet-Verlag, Schriftenr. Landesanst. Immissions- Bodennutzungssch. d. Landes Nordrhein-Westfalen **29**, 69–70 (1973)

Haselhoff, E., Bredemann, G., Haselhoff, W.: Entstehung, Erkennung und Beurteilung von Rauchschäden. Berlin: Gebr. Borntraeger Verlag, 1932

Haselhoff, E., Lindau, G.: Die Beschädigung der Vegetation durch Rauch. Leipzig: Gebr. Borntraeger Verlag, 1903

Haut, H. van: Die Analyse von Schwefeldioxidwirkungen auf Pflanzen im Laboratoriumsversuch. Staub **21**, 52–56 (1961)

Haut, H. van: Nachweis mehrerer Verunreinigungskomponenten mit Hilfe von Blätterkohl *(Brassica oleracea acephala)* als Indikatorpflanze. Staub Reinh. Luft **32**, 109–111 (1972a)

Haut, H. van: Testkammerverfahren zum Nachweis phytotoxischer Immissionskomponenten. Environ. Pollution **3**, 123–132 (1972b)

Haut, H. van, Guderian, R.: Begasungsexperimente mit Chlorwasserstoff zur Ermittlung von Dosis-Wirkungsbeziehungen bei Pflanzen. In Vorbereitung, 1976

Haut, H. van, Stratmann, H.: Experimentelle Untersuchungen über die Wirkung von Schwefeldioxid auf die Vegetation. Köln und Opladen. Westdeutscher Verlag. Forsch. Ber. d. Landes Nordrhein-Westfalen Nr. **884**, 1960

Haut, H. van, Stratmann, H.: Experimentelle Untersuchungen über die Wirkung von Stickstoffdioxid auf Pflanzen. Essen: Girardet-Verlag, Schriftenr. Landesanst. Immissions- Bodennutzungssch. d. Landes Nordrhein-Westfalen **7**, 50–70 (1967)

Haut, H. van, Stratmann, H.: Farbtafelatlas über Schwefeldioxid-Wirkungen an Pflanzen. Essen: W. Girardet-Verlag, 1970

Heagle, A. S., Body, D. E., Heck, W. W.: An open-top field chamber to assess the impact of air pollution on plants. J. of Environ. Quality **2**, 365–368 (1973)

Heath, O. V. S., Russel, J.: An investigation of the light responses of wheat stomata with the attempted elimination of control by the mesophyll. J. Exp. Botany **5**, 269–292 (1954)

Heck, W. W.: Plant injury induced by photochemical reaction products of propylene-nitrogen dioxide mixtures. JAPCA **14**, 255–261 (1964)

Heck, W. W.: The use of plants as indicators of air pollution. Intern. J. Air Water Pollution **10**, 99–111 (1966)

Heck, W. W., Dunning, J. A., Hindawi, I. J.: Interactions of environmental factors on the sensitivity of plants to air pollution. JAPCA **15**, 511–515 (1965)

Heck, W. W., Dunning, J. A., Hindawi, I. J.: Ozone: nonlinear relation of dose and injury in plants. Science **151**, 577–578 (1966)

Heggestad, H. E., Burleson, F. R., Middleton, J. T., Darley, E. F.: Leaf injury on tobacco varieties resulting from ozone, ozonated hexene-1 and ambient air of metropolitan areas. Intern. J. Air Pollution **8**, 1–10 (1964)

Heggestad, H. E., Heck, W. W.: Nature, extent and variation of plant response to air pollutants. Advan. Agron. **23**, 111–145 (1971)

Hennebo, D.: Staubfilterung durch Grünanlagen. Berlin, 1955

Heyland, K. V.: Der Verlauf der Einlagerung von Gerüstsubstanzen und anderen Kohlenhydraten in den Sproß von Weizen und Roggen zwischen Ährenschieben und Todreife. Z. Acker- Pflanzenbau **108**, 473–502 (1959)

Hilkenbäumer, F.: Obstbau, Grundlagen, Anbau und Betrieb. Berlin, Hamburg: P. Parey-Verlag, 1964

Hill, A. C.: Air quality standards for fluoride vegetation effects. JAPCA **19**, 331–336 (1969)

Hill, A. C., Transtrum, L. G., Pack, M. R., Winters, W. S.: Air pollution with relation to agronomic crops. VI. An investigation of the "hidden injury" theory of fluoride damage to plants. Agron. J. **50**, 562–565 (1958)

Hitchcock, A. E., Zimmermann, P. W., Coe, R. R.: Results of ten years work (1951–1960) on the effect of fluorides on *Gladiolus*. Contrib. Boyce Thompson Inst. **21**, 303–344 (1962)

Hitchcock, A. E., Zimmermann, P. W., Coe, R. R.: The effects of fluorides on milo maize. Contrib. Boyce Thompson Inst. **22**, 175–206 (1963)

Hodges, G. H., Menser, H. A., Ogden, W. B.: Susceptibility of Wisconsin Lavana tobacco cultivars to air pollutants. Agron. J. **63**, 107–111 (1971)

Hölte, W.: Über Fluorschäden an landwirtschaftlichen und gartenbaulichen Gewächsen durch Düngemittelfabriken. Bochum. Ber. Landesanst. Bodennutzungssch. d. Landes Nordrhein-Westfalen 42–43 (1960)

Hölte, W.: Zusammenstellung von Immissionsschadensfällen in der Zeit von 1953 bis 1972 nach Gutachten der LIB des Landes NW. Essen:. Girardet-Verlag, Schriftenr. Landesanst. Immissions- Bodennutzungssch. d. Landes Nordrhein-Westfalen **26**, 80–82 (1972)

Horsmann, D. C., Wellburn, A. R.: Synergistic effect of SO_2 and NO_2 polluted air upon enzyme activity in pea seedlings. Environ. Pollution, 123–133 (1975)

Hull, H. M., Went, F. W.: Life processes of plants as affected by air pollution. Pasadena. Proc. 2nd Nat. Poll. Symp. 122–128 (1952)

Jacobson, J. S., Hill, A. C.: Recognition of air pollution injury to vegetation: A pictorial atlas. Air Pollut. Cont. Assoc., Pittsburgh, Pa., 1970

Jacobson, J. S., Weinstein, L. H., McCune, D. C., Hitchcock, A. E.: The accumulation of fluorine by plants. JAPCA **16**, 412–417 (1966)

JAPCA: Environmental protection agency sets national air quality standards. JAPCA **21**, 352–353 (1971)

Johnson, C. M., Stout, P. R., Broyer, T. C., Carlton, A. B.: Comperative chlorine requirements of different plant species. Plant Soil **8**, 337–353 (1957)

Johnson, F., Allmendinger, D. F., Miller, V. L., Gould, C. J.: Leaf scorch of gladiolus caused by atmospheric fluoric effluents. Phytopathology **40**, 239–246 (1950)

Jones, Ruth J., Mansfield, T. A.: Increases in the diffusion resistances of leaves in a carbon dioxide-enriched atmosphere. J. Exp. Botany **21**, 951–958 (1970)

Jordan, H. V., Bardsley, C. E.: Sulfur content of rain water and atmosphere in southern states. U.S. Dep. Agri., Tech. Bull. F 96, 1–16 (1959)

Juhren, M., Noble, W. M., Wendt, F. W.: The standardization of *Poa annua* as an indicator of smog concentrations. I. Effects of temperature, photoperiod and light intensity during growth of the test-plants. Plant Physiol. **32**, 576–586 (1957)

Katz, M.: Report on the effect of dilute sulphur dioxide on alfalfa. — In: National Research Council of Canada: Trail Smelter Question. Doc. Ser. DD. Append. DD 3, Ottawa, 1937

Katz, M.: Sulphur dioxide in the atmosphere and its relation to plant life. Ind. Eng. Chem. **41** (11), 2450–2465 (1949)

Katz, M., Ledingham, G. A.: Report on the effect of dilute sulphur dioxide on alfalfa. In: National Research Council of Canada; Trail Smelter Question. Doc. Ser. DD. Append. DD 1, Ottawa, 1937

Katz, M., McCallum, A. W.: The effect of sulfur dioxide on conifers. In: Air Pollution. New York, London: Academic Press, 1952, Vol. I, pp. 401–443

Kaudy, J. D., Bingham, F. T., McColloch, R. L., Leigig, F. G., Vansklow, A. P.: Contamination of citrus foliage by fluorine from air pollution in major California citrus areas. Proc. Am. Soc. Hort. Sci. **65**, 121–127 (1955)

Keller, Th., Schwager, H.: Der Nachweis unsichtbarer („physiologischer") Fluor-Immissionsschädigungen an Waldbäumen durch eine einfache kolorimetrische Bestimmung der Peroxidase-Aktivität. Europ. J. Forest Path. **1**, 6–18 (1971)

Kendrick, J. B., Middleton, T. J., Darley, E. F.: Predisposing effects of air temperature and nitrogen supply upon injury to some herbaceous plants, fumigated with peroxides derived from olefines. Phytopathology **43**, 588 (abstracts), 1953

Kick, H.: Pflanzennährstoffe. — In: Handbuch der Pflanzenernährung und Düngung. Wien, New York: Springer-Verlag, 1969, Vol. 1/1, pp. 89–122

Kick, H., Kretzschmar, R.: Zur Anreicherung von NO_3-, SO_4-, Cl- und NH_4-Ionen im Boden und Grundwasser infolge von Düngungsmaßnahmen. Landwirtsch. Forsch. **21**, 3–18 (1968)

Kintner, R. R.: Dietary fluoride intake in the USA. Fluoride **4**, 44–48 (1971)

Kirschbaum, U., Klee, R., Steubing, L.: Flechten als Indikatoren für die Immissionsbelastung im Stadtgebiet Frankfurt/Main. Staub Reinh. Luft **31**, 21–24 (1971)

Kisser, J.: Forstliche Rauchschäden aus der Sicht des Biologen. Wien: Österr. Agrarverlag. Mitt. Forstl. Bundesversuchsanst. Mariabrunn **73**, 7–46 (1966)

Kisser, J., Bergman-Lehnert, J., Halbwachs, G.: Physiologische Ursachen charakteristischer Rauchschädigungsymptome. Wiss. Z. TU Dresden **11**, 553–559 (1962)

Klapp, E.: Wiesen und Weiden. In: Handbuch der Landwirtschaft, Pflanzenbaulehre. Berlin, Hamburg: P. Parey Verlag, 1953, Vol. II

Klapp, E.: Lehrbuch des Acker- und Pflanzenbaus. Berlin, Hamburg: P. Parey Verlag, 1967

Klapp, E., Boeker, P.: Welche Faktoren bestimmen die Höhe des Grünlandertrages? Umschau **56**, 265–267 (1956)

Knabe, W.: Experimentelle Prüfung der Fluoranreicherung in Nadeln und Blättern von Pflanzen in Abhängigkeit von deren Expositionshöhe über Grund. Zaklad Badan Naukowych Gornoslaskiego Okregu Przemyslowego Polskiej Akademii Nauk, Katowice **IX**, 9–14 (1968)

Knabe, W.: Kiefernwaldverbreitung und Schwefeldioxid-Immissionen im Ruhrgebiet. Staub Reinh. Luft **30**, 32–35 (1970a)

Knabe, W.: Natürliche Abnahme des aus den Immissionen aufgenommenen Fluors in Fichtennadeln. Kurzmitteilung. Staub Reinh. Luft **30**, 384–385 (1970b)

Knabe, W.: Immissionsbelastung und Immissionsgefährdung der Wälder im Ruhrgebiet. Essen: Girardet-Verlag, Schriftenr. Landesanst. Immissions- Bodennutzungssch. d. Landes Nordrhein-Westfalen, **26**, 83–87 (1972)

Knabe, W.: Luftverunreinigungen und Forstpflanzen. 8. Intern. Arbeitstag. forstl. Rauchschadensachverständiger 1972 in Ungarn. Allgem. Forstz. **6** (1973)

Koritz, H. G., Went, F. W.: The physiological action of smog on plants. I. Initial growth and transpiration studies. Plant Physiol. **28**, 50–62 (1953)

Koronowski, P.: Schwefel. — In: Handbuch der Pflanzenkrankheiten. Berlin, Hamburg: P. Parey Verlag, 1969, Vol. I/II, pp. 114–131

Kozel, J., Maly, V.: Agricultural production in regions exposed to industrial exhalation. Meliorace **4**, 55–63 (1968)

Krüger, E.: Die Verhütung von Rauchschäden als pflanzenzüchterisches Problem. Bergakademie, Freiberger Forschungsh. 21–23 (1951)

Kuelske, S., Prinz, B.: Untersuchungen über den Einfluß des Meßzeitraumes auf Mittelwert und Vertrauensbereich eines SO_2-Immissionswertekollektivs. Essen: Girardet-Verlag. Schriftenr. Landesanst. Immissions- u. Bodennutzungssch. d. Landes Nordrhein-Westfalen **12**, 81–91 (1968)

Kuiper, P. J. C.: Dependence upon wavelength of stomatal movement in epidermal tissue of *Senecio edoris*. Plant Physiol. **39**, 952–955 (1964)

Kurmies, B.: Über den Schwefelhaushalt des Bodens. Phosphorsäure **17**, 258–278 (1957)

Lange, O. L.: Plant water relations. Progress in Botany/Fortschritte der Botanik **37**, 78–97 (1975)
Lange, O. L., Lösch, R., Schulze, E. D., Kappen, L.: Responses of stomata to changes in humidity. Planta **100**, 76–86 (1971)
Latzko, E.: Einfluß von Cl- und SO_4-Ernährung auf die Enzymtätigkeit von Kulturpflanzen. Z. Pflanzenernähr., Düng., Bodenk. **66**, 148–155 (1954)
LeBlanc, F.: Possibilities and methods for mapping air pollution on the basis of lichen sensitivity. Wien: Agrarverlag, Mitt. Forstl. Bundesversuchsanst. Mariabrunn **92**, 103–126 (1971)
Lee, R., Gates, D. M.: Diffusion resistance in leaves as related to their stomatal anatomy and microstructure. Am. J. Botany **51**, 963–975 (1964)
Leh, H. O.: Elemente mit unzureichend geklärter Nährstoffwirkung. — In: Handbuch der Pflanzenkrankheiten. Berlin, Hamburg: P. Parey Verlag, 1969, Vol. 1/2, pp. 350–380
Leone, J. A., Brennan, E., Daines, R. H.: Atmospheric fluoride: its uptake and distribution in tomato and corn plants. Plant Physiol. **31**, 329–333 (1956)
Leone, J. A., Brennan, E.: Sulfur nutrition as it contributes to the susceptibility of tobacco and tomato to SO_2 injury. Atmos. Environ. Pergamon Press **6**, 259–266 (1972)
Levitt, J.: Responses of Plants to Environmental Stresses. New York, London: Academic Press, 1972
Lewis, R. A., Lefohn, A. S., Glass, N. R.: Introduction and perspectives. In: The Bioenvironmental Impact of a Coal-fired Power-plant. First interim report, Colstrip, Montana, Nat. Environ. Res. Center, Corvallis, 1974
Linzon, S. N.: Economic effects of sulphur dioxide on forest growth. 63rd ann. meeting of the Air Pollution Control Assn., St. Louis, Mo, June 14–18, 1970
Linzon, S. N.: Fluoride effects on vegetation in Ontario. New York. Academic Press, Proc. 2. Intern. Clean Air Congr. 277–292 (1971)
Lotfield, J. V. G.: The behaviour of stomata. Carnegie Inst. Publ. **314**, 1–104 (1921)
Lötschert, W.: Pflanzen an Grenzstandorten. Stuttgart: Fischer Verlag, 1969
Luft, J. H.: Permanganate—a new fixative for electron microscopy. J. Biophys. Biochem. Cytol. **2**, 799–801 (1956)
Lundegard, H.: Die Kohlensäureassimilation der Zuckerrübe. Flora NF **21**, 273 (1927)
MacDowall, F. D. H.: Predisposition of tobacco to ozone damage. Can. J. Plant Sci. **45**, 1–12 (1965)
MacDowall, F. D. H., Cole, A. F. W.: Threshold and synergistic damage to tobacco by ozone and sulfur dioxide. Atmos. Environ. **5**, 553–559 (1971)
MacIntire, W. H., Hardin, L. J., Buehler, M. H.: Fluorine in Maury County, Tennessee. University of Tennessee, Bull. **279**, 1–33 (1958)
MacLean, D. C., McCune, D. C., Weinstein, L. H., Mandl, R. H., Woodruff, G. N.: Effects of acute hydrogen fluoride and nitrogen dioxide exposures on citrus and ornamental plants of central Florida. Environ. Sci. Technol. **2**, 444–449 (1968)
MacLean, D. C., Schneider, R. E., Weinstein, L. H.: Accumulation of fluoride by forage crops. Contrib. Boyce Thompson Inst. **24**, 165–166 (1969)
MacLean, D. C., Schneider, R.: Fluoride phytotoxicity: Its alteration by temperature. New York, London. Proc. 2nd Intern. Clean Air Congr., 292–295 (1971)
MAGS (Minister für Arbeit, Gesundheit und Soziales des Landes Nordrhein-Westfalen): Reine Luft für morgen, Utopie oder Wirklichkeit? Möhnesee-Wamel: Verlag K. v. St.-George 1972
Malmer, N.: Gesellschaftsentwicklung und Belastung von südschwedischen Ökosystemen. Tagungsbericht der Gesellschaft für Ökologie, Tagung Gießen, 109–114 (1972)
Mandl, R. H., Weinstein, L. H., McCune, D. C., Keveny, M.: A cylindrical, open-top chamber for the exposure of plants to air pollutants in the field. J. Environ. Quality **2**, 371–376 (1973)
Manschinger, H.: Quellen der Luftverunreinigung und allgemeine Begriffsbestimmungen. In: Forstl. Rauchschäden in Österreich. Mitt. Forstl. Bundesversuchsanst. Mariabrunn. Wien: Agrarverlag, 1966, Vol. LXXIII, pp. 49–55
Mansfield, T. A., Meidner, H.: Stomatal opening in light of different wavelengths; effects of blue light independent of carbon dioxide concentration. J. Exp. Botany **17**, 510–521 (1965)

Maran, B.: Damages inflicted upon the forests in the surroundings of Krupp-Renn process plants. Sb. Cesk. Akad. Zemedel. Ved **6**, 807–818 (1960)

Marschner, H., Michael, G.: Untersuchungen über Schwefelabscheidung und Sulfataustausch an Weizenwurzeln. Z. Pflanzenernaehr. Dueng., Bodenk. **91**, 29–44 (1960)

Massey, L.M.: Similarities between disease symptoms and chemically induced injury to plants. In: Air Pollution, Proc. U.S. Tech. Congr. Air Pollution. New York: McGraw-Hill, 1952

Masuch, G., Guderian, R., Weinert, H.: Wirkung von Chlorwasserstoff auf die Ultrastruktur der Chloroplasten von *Spinacia oleracea* L. Proc. 3rd Intern. Clean Air Congr., A 160–A 163. Düsseldorf: VDI-Verlag, 1973

Materna, J.: Die Schwefeldioxideinwirkung auf die mineralische Zusammensetzung von Fichtennadeln. Arb. forstl. Forschungsanst. Tschechoslowakei **24**, 7–36 (1962)

Materna, J.: Steigerung der Widerstandsfähigkeit von Holzarten gegen Rauchgaseinwirkungen durch Düngung. Prace vyzkumnych ustavu lesnickych CSSR, svazek **26**, 209–235 (1963)

Materna, J.: Aufnahme von SO_2 durch Fichtennadeln und Weiterleitung der Schwefelverbindungen. Allgem. Forstz. **76**, 8 (1965a)

Materna, J.: Schäden an der Waldwirtschaft, die durch Exhalationen großer Wärmekraftwerke verursacht sind. Internat. Symp. über Luftreinhaltung und Verwertung von SO_2 und Flugasche aus Dampfkraftwerken. Liblice 529–536 (1965b)

Materna, J.: Die ersten Ergebnisse einer systematischen Messung von Schwefeldioxid-Immissionen im Erzgebirge. V. Internat. Arbeitstagung forstlicher Rauchschadensachverständiger, Janske Lazne, 11–14. 10., 1966

Materna, J., Jirgle, J., Kucera, J.: Vysledky mereni koncentraci kyslicniku siriciteho v lesich Krusnych hor. Ochrana ovzdusi **6**, 84–92 (1969)

Materna, J., Kohout, R.: Die Absorption des Schwefeldioxids durch die Fichte. Naturwissenschaften **50**, 407 (1963)

Materna, J., Kohout, R.: Stickstoff-Düngung und Schwefeldioxid-Aufnahme durch Fichtennadeln. Naturwissenschaften **54**, 251–252 (1967)

Matsushima, J., Brewer, R.F.: Influence of sulfur dioxide and hydorgen fluoride as a mix or reciprocal exposure on citrus growth and development. JAPCA **22**, 710–713 (1972)

McCune, D.C.: On the establishment of air quality criteria, with reference to the effects of atmospheric fluorine on vegetation. Air quality monograph 69-3. New York: Am. Petrol. Inst., 1969

McCune, D.C., Hitchcock, A.E., Jacobson, J.S., Weinstein, L.H.: Fluoride accumulation and growth of plants exposed to particulate cryolite in the atmosphere. Contrib. Boyce Thompson Inst. **23**, 1–11 (1965)

McCune, D.C., Weinstein, L.H., MacLean, D.C., Jacobson, J.S.: The concept of hidden injury in plants. Agriculture and the quality of our environment. N.C. Brady (ed.), Am. Assn. Adv. Sci., Washington D.C. 33–44 (1967)

McGlendon, J.F., Gershon-Cohen, J.: Reduction of dental caries and goiter by crops fertilized with fluorine and iodine. J. Agr. Food Chem. **3**, 72–73 (1955)

McNulty, J.B., Newman, D.W.: Mechanism of fluoride induced chlorosis. Plant Physiol. **36**, 385–388 (1961)

Meidner, H.: The comperative effect of blue and red light on the stomata of *Allium cepa* L. and *Xanthium pennsylvanicum*. J. Exp. Botany **19**, 146–151 (1968)

Meidner, H., Mansfield, T.A.: Physiology of Stomata. London: McGraw-Hill Publ. Comp., 1968

Mengel, K.: Ernährung und Stoffwechsel der Pflanze. Stuttgart: G. Fischer Verlag, 1968

Menser, H.A., Heggestad, H.E.: Ozone and sulfur dioxide synergism injury to tobacco plants. Science **153**, 424–425 (1966)

Menser, H.A., Hodges, G.H.: Effects of air pollutants on burley tobacco cultivars. Agron. J. **62**, 265–269 (1970)

Meurers, H.: Lärmminderung durch Anpflanzungen. Mitt. Forstl. Bundesversuchsanst. Wien: Wien. Österr. Agrarverlag, **97/1**, 535–539 (1972)

Middleton, J.T., Emik, L.O., Taylor, O.C.: Air quality criteria and standards for agriculture. JAPCA **15**, 476–480 (1965)

Miller, P. M., Rich, S.: Ozone damage on apples. Plant Disease. Reptr. **52**, 730–731 (1968)
Mitchell, H. H., Edman, M.: Fluorine in soils, plants and animals. Soil Sci. **60**, 81–90 (1945)
Mohr, H.: Lehrbuch der Pflanzenphysiologie. Berlin, Heidelberg, New York: Springer-Verlag, 1969
NAS: Fluorides, biologic effects of atmospheric pollutants. Nat. Acad. Sci. (Wash.) 1971
Niesslein, F.: Die Gefährdung der Flächenfunktionen des Waldes durch Rauchschäden. Wien: Österr. Agrarverlag, Mitt. Forstl. Bundesversuchsanst. Mariabrunn **73**, 213–224 (1966)
Nikfeld, H.: Pflanzensoziologische Beobachtungen im Rauchschadensgebiet eines Aluminiumwerkes. Cbl. ges. Forstwes. **84**, 318–329 (1967)
Oelschläger, W.: Fluoride in food. Fluoride **3**, 6–11 (1970)
Oelschläger, W.: Problematik der Immissionsmessungen von Fluor hinsichtlich der Wirkungen auf Pflanze und Tier. Staub Reinh. Luft **31**, 457–459 (1971)
Oertli, J. J.: Effect of salinity on susceptibility of sun flower plants to smog. Soil Sci. **87**, 249–251 (1959)
O'Gara, P. J.: Abstract of paper: Sulphur dioxide and fume problems and their solutions. Ind. Eng. Chem. **14**, 744 (1922)
Olsen, R. A.: Absorption of sulphur dioxide from the atmosphere by cotton plants. Soil Sci. **84**, 107–111 (1957)
Ottar, B.: Über die Entstehung und Folgen der sauren Niederschläge in Skandinavien. Vortrag bei VDLUFA, Kiel 1971. In Knabe, W.: Luftverunreinigung und Waldwirtschaft. Ber. üb. Landw. Hamburg: P. Parey Verlag, 1972, Vol. L, pp. 169–188
Otto, H. W., Daines, R. H.: Plant injury by air pollutants: influence of humidity on stomatal apertures and plant response to ozone. Science **163**, 1209–1210 (1969)
Pack, M. R.: Response of tomato fruiting to hydrogen fluoride as influenced by calcium nutrition. JAPCA **16**, 541–544 (1966)
Palade, G. E.: A study of fixation for electron microscopy. J. Exp. Med. **95**, 285 (1952)
Pelz, E.: Untersuchungen über die Fruktifikation rauchgeschädigter Fichtenbestände. Arch. Forstwes. **12**, 1066–1077 (1963)
Pelz, E., Materna, J.: Beiträge zum Problem der individuellen Rauchhärte von Fichte. Arch. Forstwes. **13**, (2), 177–210 (1964)
Penningsfeld, F., Forchthammer, L.: Die Düngung im Blumen- und Zierpflanzenbau. — In: Handb. der Pflanzenernährung und Düngung. Wien, New York: Springer-Verlag, 1965, pp. 917–985
Persson, G. A.: Maßnahmen zur Reinhaltung der Luft. Staub Reinh. Luft **31**, 283–284 (1971)
Peters, R. A., Murray, L. R., Shorthouse, M.: Fluoride metabolism in *Acacia georginae* Gidyea. Biochem. J. **95**, 724–730 (1965)
Peters, R. A., Shorthouse, M.: Fluoride metabolism in plants. Nature **202**, 21–22 (1964)
Peters. R. A., Wakelin, R. W., Buffa, P.: Fluoro-acetat poisoning isolation and properties of the fluorotricarboxilic acid inhibitor of citrate metabolism. Proc. Roy. Soc. **140**, 497–506 (1953)
Pfaff, C.: Einfluß der Beregnung auf die Nährstoffauswaschung bei mehrjährigem Gemüsebau. Z. Pflanzenernaehr. Dueng. Bodenk. **80**, 93–108 (1958)
Pirson, A.: Mineralstoffe und Photosynthese. In: Handbuch der Pflanzenphysiologie. Berlin, Göttingen, Heidelberg: Springer Verlag, 1958, Vol. IV, pp. 355–381
Prindle, R. A.: Air quality criteria and standards. Public Health Service Publ., 1649, 465–468 (1966)
Prinz, B., Ixfeld, H.: Schwefeldioxid-Messungen im Lande Nordrhein-Westfalen. 6. Mitteilung der Ergebnisse des III. Meßprogrammes nach § 7 des Immissionsschutzgesetzes NW für die Zeit vom 1. Nov. 1969 bis zum 31. Okt. 1970. Essen: Girardet-Verlag, Schriftenr. Landesanst. Immissions- Bodennutzungssch. d. Landes Nordrhein-Westfalen **24**, 7–52 (1971)
Prinz, B., Scholl, G.: Erhebungen über die Aufnahme und Wirkung gas- und partikelförmiger Luftverunreinigungen im Rahmen eines Wirkungskatasters. Essen: Girardet-Verlag, Schriftenr. Landesanst. Immissions- Bodennutzungssch. d. Landes Nordrhein-Westfalen **36**, 62–86 (1975)
Prinz, B., Stratmann, H.: Vorschläge zu Begriffsbestimmungen auf dem Gebiet der Luftreinhaltung. Staub Reinh. Luft **29**, 354–357 (1969)

Rao, D. N., LeBlanc, F.: Effects of sulfur dioxide in the lichen algae, with special reference to chlorophyll. Bryologist **69**, No. 1, 69–75 (1966)

Rasch, R.: Kunststoffe in der Müllverbrennung. U techn. Umweltmag. **4**, 20–23 (1971)

Reckendorfer, P.: Ein Beitrag zur Mikrochemie des Rauchschadens durch Fluor. Die Wanderung des Fluors im pflanzlichen Gewebe. I. Teil: Die unsichtbaren Schäden. Pflanzenschutzber. **9**, 33–55 (1952)

Reckendorfer, P., Beran, F.: Einfluß des Schwefeldioxides auf die Eiweißstoffe des Pflanzenorganismus. Fortschr. Landw. **6**, 435–438 (1931)

Reusmann, G., Guderian, R., Arndt, U.: Zur Wirkungsweise von Fluorwasserstoff auf höhere Pflanzen. Unveröffentlicht, 1971

Reusmann, G., Westphalen, J.: Ein elektrometrisches Verfahren zur Bestimmung des Fluorgehaltes im Pflanzenmaterial. Staub Reinh. Luft **29**, 413–415 (1969)

Reusmann, G., Westphalen, J.: Ein mechanisiertes Verfahren zur Bestimmung von Gesamtschwefel und Chlorid in Pflanzenmaterial. Essen: Girardet-Verlag, Schriftenr. Landesanst. Immissions- Bodennutzungssch. des Landes Nordrhein-Westfalen **37**, 123–128 (1976)

Reynolds, E. S.: The use of lead citrate at high pH as an electron-opaque stain in electron microscopy. J. Cell Biol. **17**, 208–212 (1963)

Riehm, H., Quellmalz, E.: Die Bestimmung der Pflanzennährstoffe im Regenwasser und in der Luft und ihre Bedeutung für die Landwirtschaft. In: Riehm, H.: Hundert Jahre staatl. Landw. Versuchs- und Forschungsanstalt Augustenberg, 1959, pp. 171–183

Rippel, A.: Fluoride intake from food. Fluoride **5**, 90–97 (1972)

Rohmeder, E., Merz, W., Schönborn, A. von: Züchtung von gegen Industrieabgase relativ resistenten Fichten- und Kiefernsorten. Forstw. Cbl. **81**, 321–332 (1962)

Rohmeder, E., Schönborn, A. von: Der Einfluß von Umwelt und Erbgut auf die Widerstandsfähigkeit der Waldbäume gegenüber Luftverunreinigungen durch Industrieabgase. Forstw. Cbl. **84**, 1–68 (1965)

Romell, L. G.: Damage of plant organs by hydrofluoric acid and other acid gases. Svensk. Bot. Tidskr. **35**, 371–378 (1941)

Rosenberger, G.: Immissionswirkungen auf Tiere. Staub Reinh. Luft **23**, 151–155 (1963)

Rosenberger, G., Gründer, H. D.: Untersuchungen über Fluorimmissionswirkungen bei Rindern im Bereich einer Flußsäurefabrik. Wiesbaden: F. Steiner-Verlag, Fluorwirkungen, Forsch. Ber. d. DFG, 68–88 (1968)

Ruge, U.: Die lichtphysiologischen Grundlagen der Pflanzenbeleuchtung. Angew. Botan. **32**, 207–220 (1958)

Saalbach, E., Judel, G. K., Kessen, G.: Über den Einfluß des Sulfatgehaltes im Boden auf die Wirkung einer Schwefeldüngung. Z. Pflanzenernaehr. Dueng. Bodenk. **99**, 177–182 (1962)

Sabatini, D. D., Bensch, K., Barnett, R. J.: Cytochemistry and electron microscopy. The preservation of cellular ultrastructure and enzymatic activity by aldehyde fixation. J. Cell Biol. **17**, 19–58 (1963)

Schade, H.: Die erwartete Schwefeldioxid-Emission aus Feuerungsanlagen in der BRD bis zum Jahre 1985 unter Berücksichtigung des Energieprogrammes 1974 der Bundesregierung. Essen: Girardet-Verlag, Schriftenr. Landesanst. Immissions- Bodennutzungssch. d. Landes Nordrhein-Westfalen **35**, 42–49 (1975)

Scheffer, T. C., Hedgcock, G. G.: Injury to northwestern forest trees by sulfur dioxide from smelters. U.S. Dep. Agri. Tech. Bull. **1117**, 1955

Schmalfuss, K.: Zur Bedeutung des Chlors als Pflanzennährstoff. Z. Pflanzenernaehr. Dueng. Bodenk. **49**, 218–223 (1950)

Schmalfuss, K.: Zur Kenntnis der Schwefelernährung der Pflanze. Z. Pflanzenernaehr. Dueng. Bodenk. **106**, 116–127 (1964)

Schmid, G.: Fluorose bei Rindern. Bull. Schweiz. Akad. med. Wiss. **12**, 397–418 (1956)

Schmidt, H., Harries, W. F., Shupe, J. L.: Fluorose bei Tieren. Schweiz. Arch. Tierheilk. **110**, 113 (1968)

Schmitz-Dumont, W.: Versuche über die Einwirkung von Fluorwasserstoff in der Atmosphäre auf Pflanzen. Tharandt. Forstl. Jb. **46**, 50–57 (1896)

Schnepf, E.: Zur Cytologie und Physiologie pflanzlicher Drüsen. Flora **153**, 1–22 (1963)

Schönbach, H., Dässler, H.-G., Enderlein, H., Bellmann, E., Kästner, W.: Über den unterschiedlichen Einfluß von Schwefeldioxid auf die Nadeln verschiedener 2jähriger Lärchenkreuzungen. Züchter **34** (8), 312–316 (1964)

Schönbeck, H.: Einfluß von Luftverunreinigungen (SO_2) auf transplantierte Flechten. Naturwissenschaften **55**, 451–452 (1968)
Schönbeck, H.: Eine Methode zur Erfassung der biologischen Wirkung von Luftverunreinigung durch transplantierte Flechten. Staub Reinh. Luft **29**, 14–18 (1969)
Schönbeck, H.: Untersuchungen in Nordrhein-Westfalen über Flechten als Indikatoren für Luftverunreinigungen. Essen: Girardet-Verlag, Schriftenr. Landesanst. Immissions- Bodennutzungssch. d. Landes Nordrhein-Westfalen **26**, 99–104 (1972)
Schönbeck, H., Buck, M., Haut, H. van, Scholl, G.: Biologische Meßverfahren für Luftverunreinigungen. Düsseldorf. VDI-Verlag, Berichte **149**, 225–236 (1970)
Schönbeck, H., Guderian, R.: Vergleichende Untersuchungen über die Reaktionen höherer Pflanzen und der Blattflechte *Hypogymnia physodes* auf gasförmige Luftverunreinigungen. In Vorbereitung, 1976
Scholl, G.: Die Immissionsrate von Fluor in Pflanzen als Maßstab für eine Immissionsbegrenzung. Düsseldorf: VDI-Verlag, Berichte **164**, 39–45 (1971a)
Scholl, G.: Ein biologisches Verfahren zur Bestimmung der Herkunft und Verbreitung von Fluorverbindungen in der Luft. Landw. Forsch. **26**, 1. Sonderheft, 29–35 (1971b)
Scholl, G.: Ein biologisches Verfahren zum Nachweis von Fluorverbindungen in Immissionen. Wien: Österr. Agrarverlag, Mitt. Forstl. Bundesversuchsanstalt Wien: **97/101**, 255–270 (1972)
Scholl, G., Schönbeck, H.: Erhebungen über Immissionsraten und Wirkungen von Luftverunreinigungen im Rahmen eines Wirkungskatasters. Essen: Girardet-Verlag, Schriftenr. Landesanst. Immissions- Bodennutzungssch. d. Landes Nordrhein-Westfalen **33**, 73–80 (1975)
Schroeder, J. von, Reuss, C.: Die Beschädigung der Vegetation durch Rauch und die Oberharzer Hüttenrauchschäden. Berlin: P. Parey Verlag 1883
Schulze, W.: Über den Einfluß der Düngung auf die Bildung der Chloroplastenpigmente. Z. Pflanzenernaehr. Dueng. Bodenk. **76**, 1–19 (1957)
Schwarz, K.: Schwefeldioxidemissionen. Staub **21**, 71–77 (1961)
Seidman, G., Hindawi, I. J., Heck, W. W.: Environmental conditions affecting the use of plants as indicators of air pollution. JAPCA **15**, 168–170 (1965)
Seidman, G., Robert, A.: Water availability and sensitivity of plants to photochemical air pollution. Plant Physiol. **38** (suppl.) XXXVI, 1963
Setterstrom, C., Zimmermann, P. W.: Apparatus for studying effects of low concentrations of gases on plants and animals. Contrib. Boyce Thompson Inst. **9**, 161–169 (1938)
Setterstrom, C., Zimmermann, P. W.: Factors influencing susceptibility of plants to sulphur dioxyde injury. Contrib. Boyce Thompson Inst. **10**, 155–181 (1939)
Shupe, J. L.: Fluorosis in livestock. Air Quality Monograph 69–74, Am. Petrol. Inst., New York, 1969
Shupe, J. L., Miner, M. L., Harris, L. E., Greenwood, D. A.: Relative effects of feeding hay atmospherically contaminated by fluoride residue, normal hay plus calcium fluoride and normal hay plus sodium fluoride to dairy heifers. Am. J. Vet. Res. **23**, 781 (1962)
Sierpinski, Z.: Einfluß von industriellen Luftverunreinigungen auf die Populationsdynamik einiger Kiefernschädlinge. München. Verh. XIV. IUFRO-Kongr., **V**, 519–531 (1967)
Skye, E.: Lichens and air pollution. Acta Phytogeographica Suecica 52, Almquist and Wiksells Boktryckeri AB, Uppsala, 1968
Slater, E. C., Bonner, W. D.: The effect of fluoride on the succinic oxidase system. Biochem. J. **52**, 185 (1952)
Solberg, P., Adams, B. F., Ferchau, H. A.: Some effects of hydrogen fluoride on the internal structure of *Pinus ponderosa* needles. Pasadena. Proc. 3rd. Nat. Air Pollut. Symp. **195**, 164–176 (1955)
Sorauer, P., Ramann, E.: Sogenannte unsichtbare Rauchbeschädigungen. Bot. Cbl. **80**, 211 (1899)
Speidel, G.: Zur Bewertung von Wohlfahrtswirkungen des Waldes. Allgem. Forstz. **21**, 383–386 (1966)
Spierings, F.: Untersuchungen von Raucheinwirkungen durch Begasungsversuche. Forschung u. Beratung, Reihe C **5**, 56–63 (1963)
Spierings, F.: Method for determining the susceptibility of trees on air pollution by artificial fumigation. Atmos. Environ. **1**, 205–210 (1967)

Spierings, F. H., Wolting, H. G.: Der Einfluß sehr niedriger HF-Konzentrationen auf die Länge der Blattspitzen-Schädigung und den Zwiebelertrag bei der Tulpenvarietät „Paris". Düsseldorf. VDI-Verlag, VDI-Berichte **164**, 19–21 (1971)

Spurr, A. R.: Refinements in the epoxy resin embedding. S. Calif. Soc. Electron. Microsc. Dec. 1–2, 1961

Stalfeld, M. G.: Temperatur. In: Handbuch d. Pflanzenphysiologie. Berlin, Göttingen, Heidelberg: Springer-Verlag, 1960, Vol. V/2, pp. 100–117

Steiner, M., Schulze-Horn, D.: Über die Verbreitung und Expositionsabhängigkeit der Rindenepiphyten im Stadtgebiet von Bonn. Bonn. Decheniana **108**, 1955

Steubing, L., Klee, R.: Vergleichende Untersuchungen zur Staubfilterwirkung von Laub- und Nadelbäumen. Angew. Botan., XLIV, 73–85 (1970)

Stöckhardt, J. A.: Über die Einwirkung des Rauches von Silberhütten auf die benachbarte Vegetation. Polyt. Centr. Bl. 257 (1850)

Stöckhardt, J. A.: Untersuchungen über die schädlichen Einwirkungen des Hütten- und Steinkohlenrauches auf das Wachstum der Pflanzen, insbesondere der Fichte und Tanne. Tharandt. Forstl. Jb. **21**, 218 ff. (1871)

Stoklasa, J.: Die Beschädigung der Vegetation durch Rauchgase und Fabrikexhalationen. München: Urban Schwartzenberg Verlag, 1923

Stout, P. R., Johnson, C. M.: Trace elements. Soil, Yearbook of Agriculture. Washington U.S.D.A. 139–150 (1957)

Stratmann, H.: Ermittlung vegetationsgefährdender SO_2-Immissionen. Landwirtsch. Forsch. **17**, 13–16 (1963a)

Stratmann, H.: Freilandversuche zur Ermittlung von Schwefeldioxidwirkungen auf die Vegetation. II. Teil: Messung und Bewertung der SO_2-Immissionen. Köln und Opladen. Westdeutscher Verlag, Forsch. Ber. d. Landes Nordrhein-Westfalen, Nr. **1184**, 1963b

Stratmann, H.: Zielsetzung im Bereich des Immissionsschutzes. Dortmund. Bergmann-Verlag, Schriftenr. d. Arbeitsgem. Rationalisierung d. Landes Nordrhein-Westfalen **133**, 1972

Stratmann, H., Buck, M., Prinz, B.: Maßstäbe für die Begrenzung der Luftverunreinigung und ihre Bedeutung. Essen: Girardet-Verlag, Schriftenr. Landesanst. Immissions- Bodennutzungssch. d. Landes Nordrhein-Westfalen **12**, 62–80 (1968)

Suttie, J. W.: Air quality standards for the protection of farm animals from fluorides. JAPCA **19**, 239–242 (1969)

Swain, R. E.: Atmospheric pollution by industrial wastes. Ind. Eng. Chem. **15**, 296–301 (1923)

TA Luft: Erste Allgemeine Verwaltungsvorschrift zum Bundes-Immissionsschutzgesetz. Technische Anleitung zur Reinhaltung der Luft vom 28. 8. 1974 (GmBl. S. 426)

Taylor, O. C.: Effects of oxidant air pollutants. J. Occupational Med. **10**, 485–492 (1968)

Templin, E.: Zur Populationsdynamik einiger Kiefernschadinsekten in rauchgeschädigten Beständen. Wiss. Z. TU Dresden **11**, 631–637 (1962)

Teworte, W.: Einsatz von fluorhaltigen Materialien in der BRD. Düsseldorf. VDI-Verlag, VDI-Berichte **164**, 11–18 (1971)

Thomas, M. D.: Gas damage to plants. Ann. Rev. Plant Physiol. **2**, 293–322 (1951)

Thomas, M. D.: Assimilation of sulphur and physiology of essential S-compounds. In: W. Ruhland: Handb. d. Pflanzenphysiol. Berlin, Göttingen, Heidelberg. Springer-Verlag **IX**, 37–63 (1958)

Thomas, M. D.: Effects of air pollution on plants. In: Air Pollution. WHO, Geneva, Monog. Series, 1961, Vol. XXXXVI, pp. 233–278

Thomas, M. D., Alther, E. W.: The effects of fluoride on plants. In: Handbuch der experimentellen Pharmakologie. Berlin, Heidelberg, New York: Springer-Verlag, 1966, Vol. XX/1, pp. 231–366

Thomas, M. D., Hendricks, R. H.: Effects of air pollution on plants. In: Air Pollution Handbook. New York: McGraw-Hill Book Comp., 1956

Thomas, M. D., Hendricks, R. H., Bryner, C. C., Hill, G. R.: A study of the sulphur metabolism of wheat, barley and corn using radioactive sulphur. Plant Physiol. **19**, 227–244 (1944a)

Thomas, M. D., Hendricks, R. H., Collier, T. R., Hill, G. R.: The utilization of sulphate and sulphur dioxide for the sulphur nutrition of alfalfa. Plant Physiol. **18**, 345–371 (1943)

Thomas, M. D., Hendricks, R. H., Hill, G. R.: Some chemical reactions of sulfur dioxide after absorption by alfalfa and sugar beets. Plant Physiol. **19**, 212–226 (1944b)

Thomas, M. D., Hendricks, R. H., Hill, G. R.: Effect of sulfur dioxide on vegetation. Ind. Eng. Chem. **42**, 2231–2235 (1950a)

Thomas, M. D., Hendricks, R. H., Hill, G. R.: Sulfur content of vegetation. Soil Sci. **70**, 9–18 (1950b)

Thomas, M. D., Hill, G. R.: Absorption of sulfur dioxide by alfalfa and it's relation to leaf injury. Plant Physiol. **10**, 291–307 (1935)

Thomas, M. D., Hill, G. R.: Relation of sulfur dioxide in the atmosphere to photosynthesis and respiration of alfalfa. Plant Physiol. **12**, 309–383 (1937)

Ting, I. P., Dugger, W. M.: Factors affecting ozone sensitivity and susceptibility of cotton plants. JAPCA **18**, 810–813 (1968)

Ting, I. P., Thompson, M. L., Dugger, W. M.: Leaf resistence to water vapor transfer in succulent plants: Effect of thermoperiod. Am. J. Botany **54**, 245–251 (1967)

Tingey, D. T., Heck, W. W., Reinert, R. A.: Effect of low concentrations of ozone and sulfur dioxide on foliage growth and yield of radish. J. Am. Soc. Hort. Sci. **96** (3), 369–371 (1971)

Tingey, D. T., Reinert, R. A., Wickliff, C., Heck, W. W.: Chronic ozone and/or sulfur dioxide exposures affect the vegetative growth of soybean. Can. J. Plant Sci. **53**, 875–879 (1973)

Tranquillini, W.: Die Bedeutung des Lichtes und der Temperatur für die Kohlensäureassimilation von *Pinus cembra* — Jungwuchs an einem hochalpinen Standort. Planta **46**, 154–178 (1955)

Trillmich, H. D.: Düngung von Mischbeständen in einem Rauchschadengebiet des Erzgebirges. Wiss. Z. TU Dresden **18**, 807–816 (1969)

Tukey, H. B., Wittwer, S. H., Bukovak, M. J.: The uptake and loss of materials by leaves and other above-ground plant parts with special reference to plant nutrition. Agrochimica **VII**, 1–28 (1962)

Türk, R., Wirth, V., Lange, O. L.: CO_2-Gaswechsel-Untersuchungen zur SO_2-Resistenz von Flechten. Oecologia **15**, 33–64 (1974)

Ulrich, A., Ohki, K.: Chlorine, bromine and sodium as nutrients for sugar beet plants. Plant Physiol. **31**, 171–181 (1956)

Ulrich, B.: Forstdüngung und Umweltschutz. Allgem. Forstz. **27**, 147–148 (1972)

VDI-Richtlinien: Maximale Immissions-Werte, VDI 2310, VDI-Handbuch Reinhaltung der Luft. Düsseldorf: VDI-Verlag, 1974

Vins, B.: A method of smoke injury evaluation-determination of increment decreace. Comm. Inst. For. Cechosloveniae 235–245 (1965)

Vins, B.: K Problematice zjistovani produkonich Ztrat v kourovych Oblastech. Lesnictvi **17**, 1033–1048 (1971)

Vogel, R.: Über die Strahlungseinflüsse auf die Stomatabewegung sowie deren Bedeutung für die Anwendung von Kunstlicht zur Pflanzenzucht. Gartenbauwissenschaft **24**, 488–526 (1960)

Vogl, M., Börtitz, S., Polster, H.: Physiologische und biochemische Beiträge zur Rauchschadenforschung. 6. Mitt. Definitionen von Schädigungsstufen und Resistenzformen gegenüber der Schadgaskomponente SO_2. Biol. Zbl. **84**, 763–777 (1965)

Walter, H., Steiner, M.: Die Ökologie der ostafrikanischen Mangroven. Z. Botan. **30**, 65–193 (1936)

Wang, T. H., Lin, C. S., Wu, C., Liao, C.: The fluorine content of Fukien tea. Food Res. **14**, 98–103 (1949)

Weinstein, L. H.: Boyce Thompson Institute for plant research, Jonkers, New York, Schreiben vom 15. 1. 1971 an den Verein Deutscher Ingenieure, Düsseldorf

Weinstein, L. H., Mandl, R. H.: The separation and collection of gaseous and particulate fluorides. Düsseldorf. VDI-Verlag, VDI-Berichte **164**, 53–71 (1971)

Weinstein, L. H., McCune, D. C.: Effects of fluoride on agriculture. JAPCA **21**, 410–413 (1971)

Wentzel, K. F.: Winterfrost 1956 und Rauchschäden. Allgem. Forstz. **11**, 541–543 (1956)

Wentzel, K. F.: Konkrete Schadwirkungen der Luftverunreinigung in der Ruhrgebietslandschaft. Natur Landschaft **37**, 118–124 (1962)

Wentzel, K. F.: Waldbauliche Maßnahmen gegen Immissionen. Allgem. Forstz. **18**, 101–106 (1963)

Wentzel, K. F.: Die Winterfrost-Schäden 1962/63 in Koniferen-Kulturen des Ruhrgebietes und ihre vermutlichen Ursachen. Forstarch. **36**, 49–59 (1965)
Wentzel, K. F.: Vorschläge zur Klassifikation der Immissionserkrankungen. Forstarch. **38**, 77–79 (1967)
Wentzel, K. F.: Empfindlichkeit und Resistenzunterschiede der Pflanzen gegenüber Luftverunreinigungen. Forstarch. **39**, 189–194 (1968)
Wiebe, H. H., Poovaiah, B. W.: Influence of moisture, heat and light stress on hydrogen fluoride fumigation injury to soy-beans. Plant Physiol. **52**, 542–545 (1973)
Wislicenus, H.: Resistenz der Fichte gegen saure Rauchgase bei ruhender und bei tätiger Assimilation. Tharandt. Forstl. Jb. **48**, 152–172 (1898)
Wislicenus, H.: Vorträge gehalten auf der Hauptversammlung des Vereins Deutscher Chemiker. Z. Angew. Chem. **28**, 689–712 (1901)
Wislicenus, H., Neger, F. W.: Experimentelle Untersuchungen über Wirkung der Abgassäuren auf die Pflanze. Berlin. Mitt. aus der Königl. Sächs. Forstl. Versuchsanst. Tharandt, **I** (3), 1914
Wittwer, S. H., Bukovac, M. J.: The uptake of nutrients through leaf surfaces. In: Handbuch der Pflanzenernäherung und Düngung. Wien, New York: Springer-Verlag Wien, 1969, Vol. 1/1, pp. 235–261
Wood, F. A.: Sources of plant pathogenic air pollutants. Phytopathology **58**, 22–31 (1968)
Zahn, R.: Wirkungen von Schwefeldioxid auf die Vegetation, Ergebnisse aus Begasungsversuchen. Staub **21**, 56–60 (1961)
Zahn, R.: Untersuchungen über die Bedeutung kontinuierlicher und intermittierender Schwefeldioxideinwirkung für die Pflanzenreaktion. Staub **23**, 343–352 (1963a)
Zahn, R.: Über den Einfluß verschiedener Umweltfaktoren auf die Pflanzenempfindlichkeit gegenüber Schwefeldioxid. Z. Pflanzenkrankh. Pflanzenschutz **70**, 81–95 (1963b)
Zattler, F., Chrometzka, P.: Untersuchungen über die Empfindlichkeit des Hopfens gegen Schwefeldioxid. Hopfen Rundschau **15**, 25–34 (1964)
Zeller, O.: Über Assimilation und Atmung der Pflanzen im Winter bei tiefen Temperaturen. Planta (Berl.) **39**, 500–526 (1951)
Zieger, E.: Rauchschäden im Walde. Wiss. Z. TH Dresden **3**, 271–280 (1953/54)
Ziegler, J.: The effect of SO_2 pollution on plant metabolism. Residue Reviews **56**, 79–105 (1975)
Zimmermann, P. W.: Impurities in the air and their influence on plant life. Proc. 1st Nat. Air Pollut. Symp. 135–141 (1950)
Zimmermann, P. W., Crocker, W.: Toxicity of air containing sulphur dioxide gas. Contrib. Boyce Thompson Inst. **6**, 455–470 (1934)
Zimmermann, P. W., Hitchcock, A. E.: Susceptibility of plants to hydrofluoric acid and sulphur dioxide gases. Contrib. Boyce Thompson Inst. **18** (6), 263–279 (1956)

Subject Index

Color Plates

Fig. 24 of Section 2.5.3.2, p. 45

Fig. 24. Degree of injury on *Ribes rubrum* leaves in various stages of development after exposure to acute SO_2 concentrations

Fig. 26. Degree of injury of *Beta vulgaris* leaves of different ages after exposure to HF. (Photo: H. van Haut)

Fig. 27 of Section 2.5.3.2, p. 47

Fig. 27. Degree of injury of *Sorbus intermedia* leaves of different ages after exposure to HCl. (Photo: H. van Haut)

Fig. 32. Effect of SO_2 on composition of a *Trifolium pratense—Lolium multiflorum* mixture. (*Left*: after exposure to SO_2; *right*: control)